Smarter Than You Think_

Smarter Than You Think_

How Technology Is Changing
Our Minds for the Better

CLIVE THOMPSON

THE PENGUIN PRESS | New York | 2013

THE PENGUIN PRESS
Published by the Penguin Group
Penguin Group (USA) LLC
375 Hudson Street
New York, New York 10014

USA | Canada | UK | Ireland | Australia | New Zealand | India | South Africa | China

penguin.com

A Penguin Random House Company

First published by The Penguin Press, a member of Penguin Group (USA) LLC, 2013

LIBRARY OF CONGRESS CATALOGING-IN-PUBLICATION DATA
Thompson, Clive.
   Smarter than you think : how technology is changing our minds for the better / Clive Thompson.
     pages cm
   Includes bibliographical references and index.
   ISBN 978-1-59420-445-6
   1. Information technology—Social aspects. 2. Internet—Social aspects.
3. Information technology—Psychological aspects. 4. Internet—Psychological aspects.
5. Social media. 6. Thought and thinking. I. Title. II. Title: How technology is changing
our minds for the better.
   T58.5.T498 2013
   303.48'33—dc23

                                2013017155

Printed in the United States of America
10 9 8 7 6 5 4 3 2

Designed by Nicole LaRoche

To Emily, Gabriel, and Zev

Contents

Smarter Than You Think_

The Rise of the Centaurs_

Who's better at chess—computers or humans?

The question has long fascinated observers, perhaps because chess seems like the ultimate display of human thought: the players sit like Rodin's *Thinker*, silent, brows furrowed, making lightning-fast calculations. It's the quintessential cognitive activity, logic as an extreme sport.

So the idea of a machine outplaying a human has always provoked both excitement and dread. In the eighteenth century, Wolfgang von Kempelen caused a stir with his clockwork Mechanical Turk—an automaton that played an eerily good game of chess, even beating Napoleon Bonaparte. The spectacle was so unsettling that onlookers cried out in astonishment when the Turk's gears first clicked into motion. But the gears, and the machine, were fake; in reality, the automaton was controlled by a chess savant cunningly tucked inside the wooden cabinet. In 1915, a Spanish inventor unveiled a genuine, honest-to-goodness robot that could actually play chess—a simple endgame involving only three pieces, anyway. A writer for *Scientific American* fretted that the inventor "Would Substitute Machinery for the Human Mind."

Eighty years later, in 1997, this intellectual standoff clanked to a dismal conclusion when world champion Garry Kasparov was defeated by IBM's Deep Blue supercomputer in a tournament of six games. Faced with a machine that could calculate two hundred million positions a second, even Kasparov's notoriously aggressive and nimble style broke down. In its final game, Deep Blue used such a clever ploy—tricking Kasparov into letting the computer sacrifice a knight—that it trounced him in nineteen moves. "I lost my fighting spirit," Kasparov said afterward, pronouncing himself "emptied completely." Riveted, the journalists announced a winner. The cover of *Newsweek* proclaimed the event "The Brain's Last Stand." Doomsayers predicted that chess itself was over. If machines could outthink even Kasparov, why would the game remain interesting? Why would anyone bother playing? What's the challenge?

Then Kasparov did something unexpected.

The truth is, Kasparov wasn't completely surprised by Deep Blue's victory. Chess grand masters had predicted for years that computers would eventually beat humans, because they understood the different ways humans and computers play. Human chess players learn by spending years studying the world's best opening moves and endgames; they play thousands of games, slowly amassing a capacious, in-brain library of which strategies triumphed and which flopped. They analyze their opponents' strengths and weaknesses, as well as their moods. When they look at the board, that knowledge manifests as intuition—a eureka moment when they suddenly spy the best possible move.

In contrast, a chess-playing computer has no intuition at all. It analyzes the game using brute force; it inspects the pieces currently on the board, then calculates all options. It prunes away moves that lead to losing positions, then takes the promising ones and runs the calculations again. After doing this a few times—and looking five or seven moves out—it arrives at a few powerful plays. The machine's way of "thinking" is fundamentally unhuman. Humans don't sit around crunching every possible move, because our brains can't hold that much information at once. If you go eight moves out in a game of chess, there are more possible games than there are stars in our galaxy. If you total up every game possible? It outnumbers the atoms in the known universe. Ask chess grand masters, "How many moves can you see out?" and they'll likely deliver the answer attributed to the Cuban grand master José Raúl Capablanca: "One, the best one."

The fight between computers and humans in chess was, as Kasparov knew, ultimately about speed. Once computers could see all games roughly seven moves out, they would wear humans down. A person might make a mistake; the computer wouldn't. Brute force wins. As he pondered Deep Blue, Kasparov mused on these different cognitive approaches.

It gave him an audacious idea. What would happen if, instead of competing against one another, humans and computers *collaborated*? What if they played on teams together—one computer and a human facing off against another human and a computer? That way, he theorized, each might benefit from the other's peculiar powers. The computer would bring the lightning-fast—if uncreative— ability to analyze zillions of moves, while the human would bring intuition and insight, the ability to read opponents and psych them out. Together, they would form what chess players later called a centaur: a hybrid beast endowed with the strengths of each.

In June 1998, Kasparov played the first public game of human-computer collaborative chess, which he dubbed "advanced chess," against Veselin Topalov, a top-rated grand master. Each used a regular computer with off-the-shelf chess software and databases of hundreds of thousands of chess games, including some of the best ever played. They considered what moves the computer recommended; they examined historical databases to see if anyone had ever been in a situation like theirs before. Then they used that information to help plan. Each game was limited to sixty minutes, so they didn't have infinite time to consult the machines; they had to work swiftly.

Kasparov found the experience "as disturbing as it was exciting." Freed from the need to rely exclusively on his memory, he was able to focus more on the creative texture of his play. It was, he realized, like learning to be a race-car driver: He had to learn how to *drive* the computer, as it were—developing a split-second sense of which strategy to enter into the computer for assessment, when to stop an unpromising line of inquiry, and when to accept or ignore the computer's advice. "Just as a good Formula One driver really knows his own car, so did we have to learn the way the computer program worked," he later wrote. Topalov, as it turns out, appeared to be an even better Formula One "thinker" than Kasparov. On purely human terms, Kasparov was a stronger player; a month before, he'd trounced Topalov 4–0. But the centaur play evened the odds. This time, Topalov fought Kasparov to a 3–3 draw.

In 2005, there was a "freestyle" chess tournament in which a team could consist of any number of humans or computers, in any combination. Many teams consisted of chess grand masters who'd won plenty of regular, human-only tournaments, achieving chess scores of 2,500 (out of 3,000). But the winning team didn't include any grand masters at all. It consisted of two young New England men, Steven Cramton and Zackary Stephen (who were comparative amateurs, with chess rankings down around 1,400 to 1,700), and their computers.

Why could these relative amateurs beat chess players with far more experience and raw talent? Because Cramton and Stephen were expert at collaborating with computers. They knew when to rely on human smarts and when to rely on the machine's advice. Working at rapid speed—these games, too, were limited to sixty minutes—they would brainstorm moves, then check to see what the computer thought, while also scouring databases to see if the strategy had occurred in previous games. They used three different computers simultaneously, running five different pieces of software; that way they could cross-check whether different programs agreed on the same move. But they wouldn't simply accept what the machine accepted, nor would they merely mimic old games. They selected moves that were low-rated by the computer if they thought they would rattle their opponents psychologically.

In essence, a new form of chess intelligence was emerging. You could rank the teams like this: (1) a chess grand master was good; (2) a chess grand master playing with a laptop was better. But even that laptop-equipped grand master could be beaten by (3) relative newbies, if the amateurs were extremely skilled at integrating machine assistance. "Human strategic guidance combined with the tactical acuity of a computer," Kasparov concluded, "was overwhelming."

Better yet, it turned out these smart amateurs could even outplay a supercomputer on the level of Deep Blue. One of the entrants that Cramton and Stephen trounced in the freestyle chess tournament was a version of Hydra, the most powerful chess computer in existence at the time; indeed, it was probably faster and stronger than Deep Blue itself. Hydra's owners let it play entirely by itself, using raw logic and speed to fight its opponents. A few days after the advanced chess event, Hydra destroyed the world's seventh-ranked grand master in a man-versus-machine chess tournament.

But Cramton and Stephen beat Hydra. They did it using their own talents and regular Dell and Hewlett-Packard computers, of the type you probably had sitting on your desk in 2005, with software you could buy for sixty dollars. All of which brings us back to our original question here: Which is smarter at chess—humans or computers?

Neither.

It's the two together, working side by side.

————

We're all playing advanced chess these days. We just haven't learned to appreciate it.

Our tools are everywhere, linked with our minds, working in tandem. Search engines answer our most obscure questions; status updates give us an ESP-like awareness of those around us; online collaborations let far-flung collaborators tackle problems too tangled for any individual. We're becoming less like Rodin's *Thinker* and more like Kasparov's centaurs. This transformation is rippling through every part of our cognition—how we learn, how we remember, and how we act upon that knowledge emotionally, intellectually, and politically. As with Cramton and Stephen, these tools can make even the amateurs among us radically smarter than we'd be on our own, assuming (and this is a big assumption) we understand how they work. At their best, today's digital tools help us see more, retain more, communicate more. At their worst, they leave us prey to the manipulation of the toolmakers. But on balance, I'd argue, what is happening is deeply positive. This book is about the transformation.

In a sense, this is an ancient story. The "extended mind" theory of cognition argues that the reason humans are so intellectually dominant is that we've always outsourced bits of cognition, using tools to scaffold our thinking into ever-more-rarefied realms. Printed books amplified our memory. Inexpensive paper and reliable pens made it possible to externalize our thoughts quickly. Studies show that our eyes zip around the page while performing long division on paper, using the handwritten digits as a form of prosthetic short-term memory. "These resources enable us to pursue manipulations and juxtapositions of ideas and data that would quickly baffle the unaugmented brain," as Andy Clark, a philosopher of the extended mind, writes.

Granted, it can be unsettling to realize how much thinking already happens outside our skulls. Culturally, we revere the Rodin ideal—the belief that genius breakthroughs come from our gray matter alone. The physicist Richard Feynman once got into an argument about this with the historian Charles Weiner. Feynman understood the extended mind; he knew that writing his equations and ideas on paper was crucial to his thought. But when Weiner looked over a pile of Feynman's notebooks, he called them a wonderful "record of his day-to-day work." No, no, Feynman replied testily. They weren't a record of his thinking process. They *were* his thinking process:

> "I actually did the work on the paper," he said.
> "Well," Weiner said, "the work was done in your head, but the record of it is still here."
> "No, it's not a *record*, not really. It's *working*. You have to work on paper and this is the paper. Okay?"

Every new tool shapes the way we think, as well as what we think about. The printed word helped make our cognition linear and abstract, along with vastly enlarging our stores of knowledge. Newspapers shrank the world; then the telegraph shrank it even more dramatically. With every innovation, cultural prophets bickered over whether we were facing a technological apocalypse or a utopia. Depending on which Victorian-age pundit you asked, the telegraph was either going usher in an era of world peace ("It is impossible that old prejudices and hostilities should longer exist," as Charles F. Briggs and Augustus Maverick intoned) or drown us in a Sargasso of idiotic trivia ("We are eager to tunnel under the Atlantic . . . but perchance the first news that will leak through into the broad, flapping American ear will be that the Princess Adelaide has the whooping cough," as Thoreau opined). Neither prediction was quite right, of course, yet neither was quite wrong. The one thing that both apocalyptics and utopians understand and agree upon is that every new technology pushes us toward new forms of behavior while nudging us away from older, familiar ones. Harold Innis—the lesser-known but arguably more interesting intellectual midwife of Marshall McLuhan—called this the bias of a new tool. Living with new technologies means understanding how they bias everyday life.

What are the central biases of today's digital tools? There are many, but I see three big ones that have a huge impact on our cognition. First, they allow for prodigious external memory: smartphones, hard drives, cameras, and sensors routinely record more information than any tool before them. We're shifting from a stance of rarely recording our ideas and the events of our lives to doing it habitually. Second, today's tools make it easier for us to find connections—between ideas, pictures, people, bits of news—that were previously invisible. Third, they encourage a superfluity of communication and publishing. This last feature has many surprising effects that are often ill understood. Any economist can tell you that when you suddenly increase the availability of a resource, people do more things with it, which also means they do increasingly unpredictable things. As electricity became cheap and ubiquitous in the West, its role expanded from things you'd expect—like nighttime lighting—to the unexpected and seemingly trivial: battery-driven toy trains, electric blenders, vibrators. The superfluity of communication today has produced everything from a rise in crowd-organized projects like Wikipedia to curious new forms of expression: television-show recaps, map-based storytelling, discussion threads that spin out of a photo posted to a smartphone app, Amazon product-review threads wittily hijacked for political satire. Now, none of these three digital biases is immutable, because they're the product of software and hardware, and can easily be altered or ended if the architects of today's tools (often corporate and governmental) decide to regulate the tools or find they're not profitable enough. But right now, these big effects dominate our current and near-term landscape.

In one sense, these three shifts—infinite memory, dot connecting, explosive publishing—are screamingly obvious to anyone who's ever used a computer. Yet they also somehow constantly surprise us by producing ever-new "tools for thought" (to use the writer Howard Rheingold's lovely phrase) that upend our mental habits in ways we never expected and often don't apprehend even as they take hold. Indeed, these phenomena have already woven themselves so deeply into the lives of people around the globe that it's difficult to stand back and take account of how much things have changed and why. While this book maps out what I call the future of thought, it's also frankly rooted in the *present*, because many parts of our future have already arrived, even if they are only dimly understood. As the sci-fi author William Gibson famously quipped: "The future is already here—it's just not very evenly distributed." This is an attempt to understand what's happening to us right now, the better to see where our augmented thought is headed. Rather than dwell in abstractions, like so many marketers and pundits—not to mention the creators of technology, who are often remarkably poor at predicting how people will use their tools—I focus more on the actual experiences of real people.

To provide a concrete example of what I'm talking about, let's take a look at something simple and immediate: my activities while writing the pages you've just read.

As I was working, I often realized I couldn't quite remember a detail and discovered that my notes were incomplete. So I'd zip over to a search engine. (*Which chess piece did Deep Blue sacrifice when it beat Kasparov? The knight!*) I also pushed some of my thinking out into the open: I blogged admiringly about the Spanish chess-playing robot from 1915, and within minutes commenters offered smart critiques. (One pointed out that the chess robot wasn't *that* impressive because it was playing an endgame that was almost impossible to lose: the robot started with a rook and a king, while the human opponent had only a mere king.) While reading Kasparov's book *How Life Imitates Chess* on my Kindle, I idly clicked on "popular highlights" to see what passages other readers had found interesting—and wound up becoming fascinated by a section on chess strategy I'd only lightly skimmed myself. To understand centaur play better, I read long, nuanced threads on chess-player discussion groups, effectively eavesdropping on conversations of people who know chess far better than I ever will. (Chess players who follow the new form of play seem divided—some think advanced chess is a grim sign of machines' taking over the game, and others think it shows that the human mind is much more valuable than computer software.) I got into a long instant-messaging session with my wife, during which I realized that I'd explained the gist of advanced chess better than I had in my original draft, so I cut and pasted that explanation into my notes. As for the act of writing itself? Like most writers, I constantly have to fight the procrastinator's urge to meander online, idly checking Twitter links and Wikipedia entries in a dreamy but pointless haze—until I look up in horror and realize I've lost two hours of work, a missing-time experience redolent of a UFO abduction. So I'd switch my word processor into full-screen mode, fading my computer desktop to black so I could see nothing but the page, giving me temporary mental peace.

In this book I explore each of these trends. First off, there's the emergence of omnipresent computer storage, which is upending the way we remember, both as individuals and as a culture. Then there's the advent of "public thinking": the ability to broadcast our ideas and the catalytic effect that has both inside and outside our minds. We're becoming more conversational thinkers—a shift that has been rocky, not least because everyday public thought uncorks the incivility and prejudices that are commonly repressed in face-to-face life. But at its best (which, I'd argue, is surprisingly often), it's a thrilling development, reigniting ancient traditions of dialogue and debate. At the same time, there's been an explosion of new forms of expression that were previously too expensive for everyday thought—like video, mapping, or data crunching. Our social awareness is shifting, too, as we develop ESP-like "ambient awareness," a persistent sense of what others are doing and thinking. On a social level, this expands our ability to understand the people we care about. On a civic level, it helps dispel traditional political problems like "pluralistic ignorance," catalyzing political action, as in the Arab Spring.

Are these changes good or bad for us? If you asked me twenty years ago, when I first started writing about technology, I'd have said "bad." In the early 1990s, I believed that as people migrated online, society's worst urges might be uncorked: pseudonymity would poison online conversation, gossip and trivia would dominate, and cultural standards would collapse. Certainly some of those predictions have come true, as anyone who's wandered into an angry political forum knows. But the truth is, while I predicted the bad stuff, I didn't foresee the good stuff. And what a torrent we have: Wikipedia, a global forest of eloquent bloggers, citizen journalism, political fact-checking—or even the way status-update tools like Twitter have produced a renaissance in witty, aphoristic, haiku-esque expression. If this book accentuates the positive, that's in part because we've been so flooded with apocalyptic warnings of late. We need a new way to talk clearly about the rewards and pleasures of our digital experiences—one that's rooted in our lived experience and also detangled from the hype of Silicon Valley.

The other thing that makes me optimistic about our cognitive future is how much it resembles our cognitive past. In the sixteenth century, humanity faced a printed-paper wave of information overload—with the explosion of books that began with the codex and went into overdrive with Gutenberg's movable type. As the historian Ann Blair notes, scholars were alarmed: How would they be able to keep on top of the flood of human expression? Who would separate the junk from what was worth keeping? The mathematician Gottfried Wilhelm Leibniz bemoaned "that horrible mass of books which keeps on growing," which would doom the quality writers to "the danger of general oblivion" and produce "a return to barbarism." Thankfully, he was wrong. Scholars quickly set about organizing the new mental environment by clipping their favorite passages from books and assembling them into huge tomes— *florilegia*, bouquets of text—so that readers could sample the best parts. They were basically blogging, going through some of the same arguments modern bloggers go through. (Is it enough to clip a passage, or do you also have to verify that what the author wrote was true? It was debated back then, as it is today.) The past turns out to be oddly reassuring, because a pattern emerges. Each time we're faced with bewildering new thinking tools, we panic—then quickly set about deducing how they can be used to help us work, meditate, and create.

History also shows that we generally improve and refine our tools to make them better. Books, for example, weren't always as well designed as they are now. In fact, the earliest ones were, by modern standards, practically unusable—often devoid of the navigational aids we now take for granted, such as indexes, paragraph breaks, or page numbers. It took decades—centuries, even—for the book to be redesigned into a more flexible cognitive tool, as suitable for quick reference as it is for deep reading. This is the same path we'll need to tread with our digital tools. It's why we need to understand not just the new abilities our tools give us today, but where they're still deficient and how they ought to improve.

I have one caveat to offer. If you were hoping to read about the neuroscience of our brains and how technology is "rewiring" them, this volume will disappoint you.

This goes against the grain of modern discourse, I realize. In recent years, people interested in how we think have become obsessed with our brain chemistry. We've marveled at the ability of brain scanning—picturing our brain's electrical activity or blood flow—to provide new clues as to what parts of the brain are linked to our behaviors. Some people panic that our brains are being deformed on a physiological level by today's technology: spend too much time flipping between windows and skimming text instead of reading a book, or interrupting your conversations to read text messages, and pretty soon you won't be able to concentrate on anything—and if you can't concentrate on it, you can't understand it either. In his book *The Shallows*, Nicholas Carr eloquently raised this alarm, arguing that the quality of our thought, as a species, rose in tandem with the ascendance of slow-moving, linear print and began declining with the arrival of the zingy, flighty Internet. "I'm not thinking the way I used to think," he worried.

I'm certain that many of these fears are warranted. It has always been difficult for us to maintain mental habits of concentration and deep thought; that's precisely why societies have engineered massive social institutions (everything from universities to book clubs and temples of worship) to encourage us to keep it up. It's part of why only a relatively small subset of people become regular, immersive readers, and part of why an even smaller subset go on to higher education. Today's multitasking tools really do make it harder than before to stay focused during long acts of reading and contemplation. They require a high level of "mindfulness"—paying attention to your own attention. While I don't dwell on the perils of distraction in this book, the importance of being mindful resonates throughout these pages. One of the great challenges of today's digital thinking tools is knowing when *not* to use them, when to rely on the powers of older and slower technologies, like paper and books.

That said, today's confident talk by pundits and journalists about our "rewired" brains has one big problem: it is very premature. Serious neuroscientists agree that we don't really know how our brains are wired to begin with. Brain chemistry is particularly mysterious when it comes to complex thought, like memory, creativity, and insight. "There will eventually be neuroscientific explanations for much of what we do; but those explanations will turn out to be incredibly complicated," as the neuroscientist Gary Marcus pointed out when critiquing the popular fascination with brain scanning. "For now, our ability to understand how all those parts relate is quite limited, sort of like trying to understand the political dynamics of Ohio from an airplane window above Cleveland." I'm not dismissing brain scanning; indeed, I'm confident it'll be crucial in unlocking these mysteries in the decades to come. But right now the field is so new that it is rash to draw conclusions, either apocalyptic or utopian, about how the Internet is changing our brains. Even Carr, the most diligent explorer in this area, cited only a single brain-scanning study that specifically probed how people's brains respond to using the Web, and those results were ambiguous.

The truth is that many healthy daily activities, if you scanned the brains of people participating in them, might appear outright dangerous to cognition. Over recent years, professor of psychiatry James Swain and teams of Yale and University of Michigan scientists scanned the brains of new mothers and fathers as they listened to recordings of their babies' cries. They found brain circuit activity similar to that in people suffering from obsessive-compulsive disorder. Now, these parents did not actually have OCD. They were just being temporarily vigilant about their newborns. But since the experiments appeared to show the brains of new parents being altered at a neural level, you could write a pretty scary headline if you wanted: BECOMING A PARENT ERODES YOUR BRAIN FUNCTION! In reality, as Swain tells me, it's much more benign. Being extra fretful and cautious around a newborn is a good thing for most parents: Babies are fragile. It's worth the tradeoff. Similarly, living in cities—with their cramped dwellings and pounding noise—stresses us out on a straightforwardly physiological level and floods our system with cortisol, as I discovered while researching stress in New York City several years ago. But the very urban density that frazzles us mentally also makes us 50 percent more productive, and more creative, too, as Edward Glaeser argues in *Triumph of the City*, because of all those connections between people. This is "the city's edge in producing ideas." The upside of creativity is tied to the downside of living in a sardine tin, or, as Glaeser puts it, "Density has costs as well as benefits." Our digital environments likely offer a similar push and pull. We tolerate their cognitive hassles and distractions for the enormous upside of being connected, in new ways, to other people.

I want to examine how technology changes our mental habits, but for now, we'll be on firmer ground if we stick to what's observably happening in the world around us: our cognitive behavior, the quality of our cultural production, and the social science that tries to measure what we do in everyday life. In any case, I won't be talking about how your brain is being "rewired." Almost everything rewires it, including this book.

The brain you had before you read this paragraph? You don't get that brain back. I'm hoping the trade-off is worth it.

———————

The rise of advanced chess didn't end the debate about man versus machine, of course. In fact, the centaur phenomenon only complicated things further for the chess world—raising questions about how reliant players were on computers and how their presence affected the game itself. Some worried that if humans got too used to consulting machines, they wouldn't be able to play without them. Indeed, in June 2011, chess master Christoph Natsidis was caught illicitly using a mobile phone during a regular human-to-human match. During tense moments, he kept vanishing for long bathroom visits; the referee, suspicious, discovered Natsidis entering moves into a piece of chess software on his smartphone. Chess had entered a phase similar to the doping scandals that have plagued baseball and cycling, except in this case the drug was software and its effect cognitive.

This is a nice metaphor for a fear that can nag at us in our everyday lives, too, as we use machines for thinking more and more. Are we losing some of our humanity? What happens if the Internet goes down: Do our brains collapse, too? Or is the question naive and irrelevant—as quaint as worrying about whether we're "dumb" because we can't compute long division without a piece of paper and a pencil?

Certainly, if we're intellectually lazy or prone to cheating and shortcuts, or if we simply don't pay much attention to how our tools affect the way we work, then yes—we can become, like Natsidis, overreliant. But the story of computers and chess offers a much more optimistic ending, too. Beause it turns out that when chess players were genuinely passionate about learning and being creative in their game, computers didn't degrade their own human abilities. Quite the opposite: it helped them internalize the game much more profoundly and advance to new levels of *human* excellence.

Before computers came along, back when Kasparov was a young boy in the 1970s in the Soviet Union, learning grand-master-level chess was a slow, arduous affair. If you showed promise and you were very lucky, you could find a local grand master to teach you. If you were one of the tiny handful who showed world-class promise, Soviet leaders would fly you to Moscow and give you access to their elite chess library, which contained laboriously transcribed paper records of the world's top games. Retrieving records was a painstaking affair; you'd contemplate a possible opening, use the catalog to locate games that began with that move, and then the librarians would retrieve records from thin files, pulling them out using long sticks resembling knitting needles. Books of chess games were rare and incomplete. By gaining access to the Soviet elite library, Kasparov and his peers developed an enormous advantage over their global rivals. That library was their cognitive augmentation.

But beginning in the 1980s, computers took over the library's role and bested it. Young chess enthusiasts could buy CD-ROMs filled with hundreds of thousands of chess games. Chess-playing software could show you how an artificial opponent would respond to any move. This dramatically increased the pace at which young chess players built up intuition. If you were sitting at lunch and had an idea for a bold new opening move, you could instantly find out which historic players had tried it, then war-game it yourself by playing against software. The iterative process of thought experiments—"If I did *this*, then what would happen?"—sped up exponentially.

Chess itself began to evolve. "Players became more creative and daring," as Frederic Friedel, the publisher of the first popular chess databases and software, tells me. Before computers, grand masters would stick to lines of attack they'd long studied and honed. Since it took weeks or months for them to research and mentally explore the ramifications of a new move, they stuck with what they knew.

But as the next generation of players emerged, Friedel was astonished by their unusual gambits, particularly in their opening moves. Chess players today, Kasparov has written, "are almost as free of dogma as the machines with which they train. Increasingly, a move isn't good or bad because it looks that way or because it hasn't been done that way before. It's simply good if it works and bad if it doesn't."

Most remarkably, it is producing players who reach grand master status younger. Before computers, it was extremely rare for teenagers to become grand masters. In 1958, Bobby Fischer stunned the world by achieving that status at fifteen. The feat was so unusual it was over three decades before the record was broken, in 1991. But by then computers had emerged, and in the years since, the record has been broken twenty times, as more and more young players became grand masters. In 2002, the Ukrainian Sergey Karjakin became one at the tender age of twelve.

So yes, when we're augmenting ourselves, we can be smarter. We're becoming centaurs. But our digital tools can also leave us smarter even when we're not actively using them.

Let's turn to a profound area where our thinking is being augmented: the world of infinite memory.

We, the Memorious_

What prompts a baby, sitting on the kitchen floor at eleven months old, to suddenly blurt out the word "milk" for the first time? Had the parents said the word more frequently than normal? How many times had the baby heard the word pronounced—three thousand times? Or four thousand times or ten thousand? Precisely how long does it take before a word sinks in anyway? Over the years, linguists have tried to ask parents to keep diaries of what they say to their kids, but it's ridiculously hard to monitor household conversation. The parents will skip a day or forget the details or simply get tired of the process. We aren't good at recording our lives in precise detail, because, of course, we're busy living them.

In 2005, MIT speech scientist Deb Roy and his wife, Rupal Patel (also a speech scientist) were expecting their first child—a golden opportunity, they realized, to observe the boy developing language. But they wanted to do it scientifically. They wanted to collect an actual record of every single thing they, or *anyone*, said to the child—and they knew it would work only if the recording was done automatically. So Roy and his MIT students designed "TotalRecall," an audacious setup that involved wiring his house with cameras and microphones. "We wanted to create," he tells me, "the ultimate memory machine."

In the months before his son arrived, Roy's team installed wide-angle video cameras and ultrasensitive microphones in every room in his house. The array of sensors would catch every interaction "down to the whisper" and save it on a huge rack of hard drives stored in the basement. When Roy and his wife brought their new-born home from the hospital, they turned the system on. It began producing a firehose of audio and video: About 300 gigabytes per day, or enough to fill a normal laptop every twenty-four hours. They kept it up for two years, assembling a team of grad students and scientists to analyze the flow, transcribe the chatter, and figure out how, precisely, their son learned to speak.

They made remarkable discoveries. For example, they found that the boy had a burst of vocabulary acquisition—"word births"—that began around his first birthday and then slowed drastically seven months later. When one of Roy's grad students analyzed this slow-down, an interesting picture emerged: At the precise moment that those word births were decreasing, the boy suddenly began using far more two-word sentences. "It's as if he shifted his cognitive effort from learning new words to generating novel sentences," as Roy later wrote about it. Another grad student discovered that the boy's caregivers tended to use certain words in specific locations in the house—the word "don't," for example, was used frequently in the hallway, possibly because caregivers often said "don't play on the stairs." And location turned out to be important: The boy tended to learn words more quickly when they were linked to a particular space. It's a tantalizing finding, Roy points out, because it suggests we could help children learn language more effectively by changing where we use words around them. The data is still being analyzed, but his remarkable experiment has the potential to transform how early-language acquisition is understood.

It has also, in an unexpected way, transformed Roy's *personal* life. It turns out that by creating an insanely nuanced scientific record of his son's first two years, Roy has created the most detailed memoir in history.

For example, he's got a record of the first day his son walked.

On-screen, you can see Roy step out of the bathroom and notice the boy standing, with a pre-toddler's wobbly balance, about six feet away. Roy holds out his arms and encourages him to walk over: "Come on, come on, you can do it," he urges. His son lurches forward one step, then another, and another—his first time successfully doing this. On the audio, you can actually hear the boy squeak to himself in surprise: *Wow!* Roy hollers to his mother, who's visiting and is in the kitchen: "He's walking! He's walking!"

It's rare to catch this moment on video for any parent. But there's something even more unusual about catching it *unintentionally.* Unlike most first-step videos caught by a camera-phone-equipped parent, Roy wasn't actively trying to freeze this moment; he didn't get caught up in the strange, quintessentially modern dilemma that comes from trying to simultaneously experience something delightful while also acting and getting it on tape. (When we brought my son a candle-bedecked cupcake on his first birthday, I spent so much time futzing with snapshots—it turns out cheap cameras don't focus well when the lights are turned off—that I later realized I hadn't actually watched the moment with my own eyes.) You can see Roy genuinely lost in the moment, enthralled. Indeed, he only realized weeks after his son walked that he could hunt down the digital copy; when he pulled it out, he was surprised to find he'd completely misremembered the event. "I originally remembered it being a sunny morning, my wife in the kitchen," he says. "And when we finally got the video it was not a sunny morning, it was evening; and it was not my wife in the kitchen, it was my mother."

Roy can perform even crazier feats of recall. His system is able to stitch together the various video streams into a 3-D view. This allows you to effectively "fly" around a recording, as if you were inside a video game. You can freeze a moment, watch it backward, all while flying through; it's like a TiVo for reality. He zooms into the scene of his watching his son, freezes it, then flies down the hallway into the kitchen, where his mother is looking up, startled, reacting to his yells of delight. It seems wildly futuristic, but Roy claims that eventually it won't be impossible to do in your own home: cameras and hard drives are getting cheaper and cheaper, and the software isn't far off either.

Still, as Roy acknowledges, the whole project is unsettling to some observers. "A lot of people have asked me, 'Are you insane?'" He chuckles. They regard the cameras as Orwellian, though this isn't really accurate; it's Roy who's recording himself, not a government or evil corporation, after all. But still, wouldn't living with incessant recording corrode daily life, making you afraid that your weakest moments—bickering mean-spiritedly with your spouse about the dishes, losing your temper over something stupid, or, frankly, even having sex—would be recorded forever? Roy and his wife say this didn't happen, because they were in control of the system. In each room there was a control panel that let you turn off the camera or audio; in general, they turned things off at 10 p.m. (after the baby was in bed) and back on at 8 a.m. They also had an "oops" button in every room: hit it, and you could erase as much as you wanted from recent recordings—a few minutes, an hour, even a day. It was a neat compromise, because of course one often doesn't know when something embarrassing is going to happen until it's already happening.

"This came up from, you know, my wife breast-feeding," Roy says. "Or I'd stumble out of the shower, dripping and naked, wander out in the hallway—then realize what I was doing and hit the 'oops' button. I didn't think my grad students needed to see *that*." He also experienced the effect that documentarians and reality TV producers have long noticed: after a while, the cameras vanish.

The downsides, in other words, were worth the upsides—both scientific and personal. In 2007, Roy's father came over to see his grandson when Roy was away at work. A few months later, his father had a stroke and died suddenly. Roy was devastated; he'd known his father's health was in bad shape but hadn't expected the end to come so soon.

Months later, Roy realized that he'd missed the chance to see his father play with his grandson for the last time. But the house had autorecorded it. Roy went to the TotalRecall system and found the video stream. He pulled it up: his father stood in the living room, lifting his grandson, tickling him, cooing over how much he'd grown.

Roy froze the moment and slowly panned out, looking at the scene, rewinding it and watching again, drifting around to relive it from several angles.

"I was floating around like a ghost watching him," he says.

What would it be like to never forget anything? To start off your life with that sort of record, then keep it going until you die?

Memory is one of the most crucial and mysterious parts of our identities; take it away, and identity goes away, too, as families wrestling with Alzheimer's quickly discover. Marcel Proust regarded the recollection of your life as a defining task of humanity; meditating on what you've done is an act of recovering, literally hunting around for "lost time." Vladimir Nabokov saw it a bit differently: in *Speak, Memory*, he sees his past actions as being so deeply intertwined with his present ones that he declares, "I confess I do not believe in time." (As Faulkner put it, "The past is never dead. It's not even past.")

In recent years, I've noticed modern culture—in the United States, anyway—becoming increasingly, almost frenetically obsessed with lapses of memory. This may be because the aging baby-boomer population is skidding into its sixties, when forgetting the location of your keys becomes a daily embarrassment. Newspaper health sections deliver panicked articles about memory loss and proffer remedies, ranging from advice that is scientifically solid (get more sleep and exercise) to sketchy (take herbal supplements like ginkgo) to corporate snake oil (play pleasant but probably useless "brain fitness" video games.) We're pretty hard on ourselves. Frailties in memory are seen as frailties in intelligence itself. In the run-up to the American presidential election of 2012, the candidacy of a prominent hopeful, Rick Perry, began unraveling with a single, searing memory lapse: in a televised debate, when he was asked about the three government bureaus he'd repeatedly vowed to eliminate, Perry named the first two—but was suddenly unable to recall the third. He stood there onstage, hemming and hawing for fifty-three agonizing seconds before the astonished audience, while his horrified political advisers watched his candidacy implode. ("It's over, isn't it?" one of Perry's donors asked.)

Yet the truth is, the politician's mishap wasn't all that unusual. On the contrary, it was extremely normal. Our brains are remarkably bad at remembering details. They're great at getting the gist of something, but they consistently muff the specifics. Whenever we read a book or watch a TV show or wander down the street, we extract the meaning of what we see—the parts of it that make sense to us and fit into our overall picture of the world—but we lose everything else, in particular discarding the details that don't fit our predetermined biases. This sounds like a recipe for disaster, but scientists point out that there's an upside to this faulty recall. If we remembered every single detail of everything, we wouldn't be able to make sense of anything. Forgetting is a gift and a curse: by chipping away at what we experience in everyday life, we leave behind a sculpture that's meaningful to us, even if sometimes it happens to be wrong.

Our first glimpse into the way we forget came in the 1880s, when German psychologist Hermann Ebbinghaus ran a long, fascinating experiment on himself. He created twenty-three hundred "nonsense" three-letter combinations and memorized them. Then he'd test himself at regular periods to see how many he could remember. He discovered that memory decays quickly after you've learned something: Within twenty minutes, he could remember only about 60 percent of what he'd tried to memorize, and within an hour he could recall just under a half. A day later it had dwindled to about one third. But then the pace of forgetting slowed down. Six days later the total had slipped just a bit more—to 25.4 percent of the material—and a month later it was only a little worse, at 21.1 percent. Essentially, he had lost the great majority of the three-word combinations, but the few that remained had passed into long-term memory. This is now known as the Ebbinghaus curve of forgetting, and it's a good-news-bad-news story: Not much gets into long-term memory, but what gets there sticks around.

Ebbinghaus had set himself an incredibly hard memory task. Meaningless gibberish is by nature hard to remember. In the 1970s and '80s, psychologist Willem Wagenaar tried something a bit more true to life. Once a day for six years, he recorded a few of the things that happened to him on notecards, including details like where it happened and who he was with. (On September 10, 1983, for example, he went to see Leonardo da Vinci's *Last Supper* in Milan with his friend Elizabeth Loftus, the noted psychologist). This is what psychologists call "episodic" or "autobiographical" memory— things that happen to us personally. Toward the end of the experiment, Wagenaar tested himself by pulling out a card to see if he remembered the event. He discovered that these episodic memories don't degrade anywhere near as quickly as random information: In fact, he was able to recall about 70 percent of the events that had happened a half year ago, and his memory gradually dropped to 29 percent for events five years old. Why did he do better than Ebbinghaus? Because the cards contained "cues" that helped jog his memory—like knowing that his friend Liz Loftus was with him—and because some of the events were inherently more memorable. Your ability to recall something is highly dependent on the context in which you're trying to do so; if you have the right cues around, it gets easier. More important, Wagenaar also showed that committing something to memory in the first place is much simpler if you're paying close attention. If you're engrossed in an emotionally vivid visit to a da Vinci painting, you're far more likely to recall it; your everyday humdrum Monday meeting, not so much. (And if you're frantically multitasking on a computer, paying only partial attention to a dozen tasks, you might only dimly remember *any* of what you're doing, a problem that I'll talk about many times in this book.) But even so, as Wagenaar found, there are surprising limits. For fully 20 percent of the events he recorded, he couldn't remember anything at all.

Even when we're able to remember an event, it's not clear we're remembering it correctly. Memory isn't passive; it's active. It's not like pulling a sheet from a filing cabinet and retrieving a precise copy of the event. You're also *regenerating* the memory on the fly. You pull up the accurate gist, but you're missing a lot of details. So you imaginatively fill in the missing details with stuff that seems plausible, whether or not it's actually what happened. There's a reason why we call it "re-membering"; we reassemble the past like Frankenstein assembling a body out of parts. That's why Deb Roy was so stunned to look into his TotalRecall system and realize that he'd mentally mangled the details of his son's first steps. In reality, Roy's mother was in the kitchen and the sun was down—but Roy remembered it as his wife being in the kitchen on a sunny morning. As a piece of narrative, it's perfectly understandable. The memory feels much more magical that way: The sun shining! The boy's mother nearby! Our minds are drawn to what *feels* true, not what's necessarily so. And worse, these filled-in errors may actually compound over time. Some memory scientists suspect that when we misrecall something, we can store the false details in our memory in what's known as reconsolidation. So the next time we remember it, we're pulling up false details; maybe we're even adding new errors with each act of recall. Episodic memory becomes a game of telephone played with oneself.

The malleability of memory helps explain why, over decades, we can adopt a surprisingly rewritten account of our lives. In 1962, the psychologist Daniel Offer asked a group of fourteen-year-old boys questions about significant aspects of their lives. When he hunted them down thirty-four years later and asked them to think back on their teenage years and answer precisely the same questions, their answers were remarkably different. As teenagers, 70 percent said religion was helpful to them; in their forties, only 26 percent recalled that. Fully 82 percent of the teenagers said their parents used corporal punishment, but three decades later, only one third recalled their parents hitting them. Over time, the men had slowly revised their memories, changing them to suit the ongoing shifts in their personalities, or what's called hindsight bias. If you become less religious as an adult, you might start thinking that's how you were as a child, too.

For eons, people have fought back against the fabrications of memory by using external aids. We've used chronological diaries for at least two millennia, and every new technological medium increases the number of things we capture: George Eastman's inexpensive Brownie camera gave birth to everyday photography, and VHS tape did the same thing for personal videos in the 1980s. In the last decade, though, the sheer welter of artificial memory devices has exploded, so there are more tools capturing shards of our lives than ever before—e-mail, text messages, camera phone photos and videos, note-taking apps and word processing, GPS traces, comments, and innumerable status updates. (And those are just the *voluntary* recordings you participate in. There are now innumerable government and corporate surveillance cameras recording you, too.)

The biggest shift is that most of this doesn't require much work. Saving artificial memories used to require foresight and effort, which is why only a small fraction of very committed people kept good diaries. But digital memory is frequently passive. You don't *intend* to keep all your text messages, but if you've got a smartphone, odds are they're all there, backed up every time you dock your phone. Dashboard cams on Russian cars are supposed to help drivers prove their innocence in car accidents, but because they're always on, they also wound up recording a massive meteorite entering the atmosphere. Meanwhile, today's free e-mail services like Gmail are biased toward permanent storage; they offer such capacious memory that it's easier for the user to keep everything than to engage in the mental effort of deciding whether to delete each individual message. (This is an intentional design decision on Google's part, of course; the more they can convince us to retain e-mail, the more data about our behavior they have in order to target ads at us more effectively.) And when people buy new computers, they rarely delete old files—in fact, research shows that most of us just copy our old hard drives onto our new computers, and do so again three years later with our next computers, and on and on, our digital external memories nested inside one other like wooden dolls. The cost of storage has plummeted so dramatically that it's almost comical to consider: In 1981, a gigabyte of memory cost roughly three hundred thousand dollars, but now it can be had for pennies.

We face an intriguing inversion point in human memory. We're moving from a period in which most of the details of our lives were forgotten to one in which many, perhaps most of them, will be captured. How will that change the way we live—and the way we understand the shape of our lives?

There's a small community of people who've been trying to figure this out by recording as many bits of their lives as they can as often as possible. They don't want to lose a detail; they're trying to create perfect recall, to find out what it's like. They're the lifeloggers.

When I interview someone, I take pretty obsessive notes: not only everything they say, but also what they look like, how they talk. Within a few minutes of meeting Gordon Bell, I realized I'd met my match: His digital records of me were thousands of times more complete than my notes about him.

Bell is probably the world's most ambitious and committed lifelogger. A tall and genial white-haired seventy-eight-year-old, he walks around outfitted with a small fish-eye camera hanging around his neck, snapping pictures every sixty seconds, and a tiny audio recorder that captures most conversations. Software on his computer saves a copy of every Web page he looks at and every e-mail he sends or receives, even a recording of every phone call.

"Which is probably illegal, but what the hell," he says with a guffaw. "I never know what I'm going to need later on, so I keep everything." When I visited him at his cramped office in San Francisco, it wasn't the first time we'd met; we'd been hanging out and talking for a few days. He typed "Clive Thompson" into his desktop computer to give me a taste of what his "surrogate brain," as he calls it, had captured of me. (He keeps a copy of his lifelog on his desktop and his laptop.) The screen fills with a flood of Clive-related material: twenty-odd e-mails Bell and I had traded, copies of my articles he'd perused online, and pictures beginning with our very first meeting, a candid shot of me with my hand outstretched. He clicks on an audio file from a conversation we'd had the day before, and the office fills with the sound of the two of us talking about a jazz concert he'd seen in Australia with his wife. It's eerie hearing your own voice preserved in somebody else's memory base. Then I realize in shock that when he'd first told me that story, I'd taken down incorrect notes about it. I'd written that he was with his daughter, not his wife. Bell's artificial memory was correcting *my* memory.

Bell did not intend to be a pioneer in recording his life. Indeed, he stumbled into it. It started with a simple desire: He wanted to get rid of stacks of paper. Bell has a storied history; in his twenties, he designed computers, back when they were the size of refrigerators, with spinning hard disks the size of tires. He quickly became wealthy, quit his job to become a serial investor, and then in the 1990s was hired by Microsoft as an éminence grise, tasked with doing something vaguely futuristic—whatever he wanted, really. By that time, Bell was old enough to have amassed four filing cabinets crammed with personal archives, ranging from programming memos to handwritten letters from his kid and weird paraphernalia like a "robot driver's license." He was sick of lugging it around, so in 1997 he bought a scanner to see if he could go paperless. Pretty soon he'd turned a lifetime of paper into searchable PDFs and was finding it incredibly useful. So he started thinking: Why not have a copy of *everything* he did? Microsoft engineers helped outfit his computer with autorecording software. A British engineer showed him the SenseCam she'd invented. He began wearing that, too. (Except for the days where he's worried it'll stop his heart. "I've been a little leery of wearing it for the last week or so because the pacemaker company sent a little note around," he tells me. He had a massive heart attack a few years back and had a pacemaker implanted. "Pacemakers don't like magnets, and the SenseCam has one." One part of his cyborg body isn't compatible with the other.)

The truth is, Bell looks a little nuts walking around with his recording gear strapped on. He knows this; he doesn't mind. Indeed, Bell possesses the dry air of a wealthy older man who long ago ceased to care what anyone thinks about him, which is probably why he was willing to make his life into a radical experiment. He also, frankly, seems like someone who *needs* an artificial memory, because I've rarely met anyone who seems so scatterbrained in everyday life. He'll start talking about one subject, veer off to another in midsentence, only to interrupt that sentence with another digression. If he were a teenager, he'd probably be medicated for ADD.

Yet his lifelog does indeed let him perform remarkable memory feats. When a friend has a birthday, he'll root around in old handwritten letters to find anecdotes for a toast. For a commencement address, he dimly recalled a terrific aphorism that he'd pinned to a card above his desk three decades before, and found it: "Start many fires." Given that he's old, his health records have become quite useful: He's used SenseCam pictures of his post-heart-attack chest rashes to figure out whether he was healing or not, by quickly riffling through them like a flip-book. "Doctors are always asking you stuff like 'When did this pain begin?' or 'What were you eating on such and such a day?'—and that's precisely the stuff we're terrible at remembering," he notes. While working on a Department of Energy task force a few years ago, he settled an argument by checking the audio record of a conference call. When he tried to describe another jazz performance, he found himself tongue-tied, so he just punched up the audio and played it.

Being around Bell is like hanging out with some sort of mnemonic performing seal. I wound up barking weird trivia questions just to see if he could answer them. When was the first-ever e-mail you sent your son? *1996.* Where did you go to church when you were a kid? *Here's a First Methodist Sunday School certificate.* Did you leave a tip when you bought a coffee this morning on the way to work? *Yep—here's the pictures from Peet's Coffee.*

But Bell believes the deepest effects of his experiment aren't just about being able to recall details of his life. I'd expected him to be tied to his computer umbilically, pinging it to call up bits of info all the time. In reality, he tends to consult it sparingly—mostly when I prompt him for details he can't readily bring to mind.

The long-term effect has been more profound than any individual act of recall. The lifelog, he argues, given him greater mental peace. Knowing there's a permanent backup of almost everything he reads, sees, or hears allows him to live more in the moment, paying closer attention to what he's doing. The anxiety of committing something to memory is gone.

"It's a freeing feeling," he says. "The fact that I can offload my memory, knowing that it's there—that whatever I've seen can be found again. I feel cleaner, lighter."

The problem is that while Bell's offboard memory may be immaculate and detailed, it can be curiously hard to search. Your organic brain may contain mistaken memories, but generally it finds things instantaneously and fluidly, and it's superb at flitting from association to association. If we had met at a party last month and you're now struggling to remember my name, you'll often sift sideways through various cues—who else was there? what were we talking about? what music was playing?—until one of them clicks, and *ping*: The name comes to us. (Clive Thompson!) In contrast, digital tools don't have our brain's problem with inaccuracy; if you give it "Clive," it'll quickly pull up everything with a "Clive" associated, in perfect fidelity. But machine searching is brittle. If you don't have the right cue to start with—say, the name "Clive"—or if the data didn't get saved in the right way, you might never find your way back to my name.

Bell struggles with these machine limits all the time. While eating lunch in San Francisco, he tells me about a Paul Krugman column he liked, so I ask him to show it to me. But he can't find it on the desktop copy of his lifelog: His search for "Paul Krugman" produces scores of columns, and Bell can't quite filter out the right one. When I ask him to locate a colleague's phone number, he runs into another wall: he can locate all sorts of things—even audio of their last conversation—but no number. "Where the hell is this friggin' phone call?" he mutters, pecking at the keyboard. "I either get nothing or I get too much!" It's like a scene from a Philip K. Dick novel: A man has external memory, but it's locked up tight and he can't access it—a cyborg estranged from his own mind.

As I talked to other lifeloggers, they bemoaned the same problem. Saving is easy; finding can be hard. Google and other search engines have spent decades figuring out how to help people find things on the Web, of course. But a Web search is actually *easier* than searching through someone's private digital memories. That's because the Web is filled with social markers that help Google try to guess what's going to be useful. Google's famous PageRank system looks at social rankings: If a Web page has been linked to by hundreds of other sites, Google guesses that that page is important in some way. But lifelogs don't have that sort of social data; unlike blogs or online social networks, they're a private record used only by you.

Without a way to find or make sense of the material, a lifelog's greatest strength—its byzantine, brain-busting level of detail—becomes, paradoxically, its greatest flaw. Sure, go ahead and archive your every waking moment, but how do you parse it? Review it? Inspect it? Nobody has *another* life in which to relive their previous one. The lifelogs remind me of Jorge Luis Borges's story "On Exactitude in Science," in which a group of cartographers decide to draw a map of their empire with a 1:1 ratio: it is the exact size of the actual empire, with the exact same detail. The next generation realizes that a map like that is useless, so they let it decay. Even if we are moving toward a world where less is forgotten, that isn't the same as more being *remembered*.

Cathal Gurrin probably has the most heavily photographed life in history, even more than Bell. Gurrin, a researcher at Dublin City University, began wearing a SenseCam five years ago and has ten million pictures. The SenseCam has preserved candid moments he'd never otherwise have bothered to shoot: the time he lounged with friends in his empty house the day before he moved; his first visit to China, where the SenseCam inadvertently captured the last-ever pictures of historic buildings before they were demolished in China's relentless urban construction upheaval. He's dipped into his log to try to squirm out of a speeding ticket (only to have his SenseCam prove the police officer was right; another self-serving memory distortion on the part of his organic memory).

But Gurrin, too, has found that it can be surprisingly hard to locate a specific image. In a study at his lab, he listed fifty of his "most memorable" moments from the last two and a half years, like his first encounters with new friends, last encounters with loved ones, and meeting TV celebrities. Then, over the next year and a half, his labmates tested him to see how quickly he could find a picture of one of those moments. The experiment was gruesome: The first searches took over thirteen minutes. As the lab slowly improved the image-search tools, his time dropped to about two minutes, "which is still pretty slow," as one of his labmates noted. This isn't a problem just for lifeloggers; even middle-of-the-road camera phone users quickly amass so many photos that they often give up on organizing them. Steve Whittaker, a psychologist who designs interfaces and studies how we interact with computers, asked a group of subjects to find a personally significant picture on their own hard drive. Many couldn't. "And they'd get pretty upset when they realized that stuff was there, but essentially gone," Whittaker tells me. "We'd have to reassure them that 'no, no, everyone has this problem!'" Even Gurrin admits to me that he rarely searches for anything at all in his massive archive. He's waiting for better search tools to emerge.

Mind you, he's confident they will. As he points out, fifteen years ago you couldn't find much on the Web because the search engines were dreadful. "And the first MP3 players were horrendous for finding songs," he adds. The most promising trends in search algorithms include everything from "sentiment analysis" (you could hunt for a memory based on how happy or sad it is) to sophisticated ways of analyzing pictures, many of which are already emerging in everyday life: detecting faces and locations or snippets of text in pictures, allowing you to hunt down hard-to-track images by starting with a vague piece of half recall, the way we interrogate our own minds. The app Evernote has already become popular because of its ability to search for text, even bent or sideways, within photos and documents.

Yet the weird truth is that *searching* a lifelog may not, in the end, be the way we take advantage of our rapidly expanding artificial memory. That's because, ironically, searching for something leaves our imperfect, gray-matter brain in control. Bell and Gurrin and other lifeloggers have superb records, but they don't search them unless, while using their own brains, they realize there's something to look for. And of course, our organic brains are riddled with memory flaws. Bell's lifelog could well contain the details of a great business idea he had in 1992; but if he's forgotten he ever had that idea, he's unlikely to search for it. It remains as remote and unused as if he'd never recorded it at all.

The real promise of artificial memory isn't its use as a passive storage device, like a pen-and-paper diary. Instead, future lifelogs are liable to be active—trying to remember things *for* us. Lifelogs will be far more useful when they harness what computers are uniquely good at: brute-force pattern finding. They can help us make sense of our archives by finding connections and reminding us of what we've forgotten. Like the hybrid chess-playing centaurs, the solution is to let the computers do what they do best while letting humans do what they do best.

Bradley Rhodes has had a taste of what that feels like. While a student at MIT, he developed the Remembrance Agent, a piece of software that performed one simple task. The agent would observe what he was typing—e-mails, notes, an essay, whatever. It would take the words he wrote and quietly scour through years of archived e-mails and documents to see if anything he'd written in the past was similar in content to what he was writing about now. Then it would offer up snippets in the corner of the screen—close enough for Rhodes to glance at.

Sometimes the suggestions were off topic and irrelevant, and Rhodes would ignore them. But frequently the agent would find something useful—a document Rhodes had written but forgotten about. For example, he'd find himself typing an e-mail to a friend, asking how to work the campus printer, when the agent would show him that he already had a document that contained the answer. Another time, Rhodes—an organizer for MIT's ballroom dance club—got an e-mail from a club member asking when the next event was taking place. Rhodes was busy with schoolwork and tempted to blow him off, but the agent pointed out that the club member had asked the same question a month earlier, and Rhodes hadn't answered then either.

"I realized I had to switch gears and apologize and go, 'Sorry for not getting back to you,'" he tells me. The agent wound up saving him from precisely the same spaced-out forgetfulness that causes us so many problems, interpersonal and intellectual, in everyday life. "It keeps you from looking stupid," he adds. "You discover things even you didn't know you knew." Fellow students started pestering him for trivia. "They'd say, 'Hey Brad, I know you've got this augmented brain, can you answer this?'"

In essence, Rhodes's agent took advantage of computers' sheer tirelessness. Rhodes, like most of us, isn't going to bother running a search on *everything* he has ever typed on the off chance that it might bring up something useful. While machines have no problem doing this sort of dumb task, they won't know if they've found anything useful; it's up to us, with our uniquely human ability to recognize useful information, to make that decision. Rhodes neatly hybridized the human skill at creating meaning with the computer's skill at making connections.

Granted, this sort of system can easily become too complicated for its own good. Microsoft is still living down its disastrous introduction of Clippy, a ghastly piece of artificial intelligence—I'm using that term very loosely—that would observe people's behavior as they worked on a document and try to bust in, offering "advice" that tended to be spectacularly useless.

The way machines will become integrated into our remembering is likely to be in smaller, less intrusive bursts. In fact, when it comes to finding meaning in our digital memories, less may be more. Jonathan Wegener, a young computer designer who lives in Brooklyn, recently became interested in the extensive data trails that he and his friends were leaving in everyday life: everything from Facebook status updates to text messages to blog posts and check-ins at local bars using services like Foursquare. The check-ins struck him as particularly interesting. They were geographic; if you picked a day and mapped your check-ins, you'd see a version of yourself moving around the city. It reminded him of a trope from the video games he'd played as a kid: "racing your ghost." In games like *Mario Kart*, if you had no one to play with, you could record yourself going as fast as you could around a track, then compete against the "ghost" of your former self.

Wegener thought it would be fun to do the same thing with check-ins—show people what they'd been doing on a day in their past. In one hectic weekend of programming, he created a service playfully called FoursquareAnd7YearsAgo. Each day, the service logged into your Foursquare account, found your check-ins from one year back (as well as any "shout" status statements you made), and e-mailed a summary to you. Users quickly found the daily e-mail would stimulate powerful, unexpected bouts of reminiscence. I spent an afternoon talking to Daniel Giovanni, a young social-media specialist in Jakarta who'd become a mesmerized user of FoursquareAnd7YearsAgo. The day we spoke was the one-year anniversary of his thesis defense, and as he looked at the list of check-ins, the memories flooded back: at 7:42 a.m. he showed up on campus to set up (with music from *Transformers 2* pounding in his head, as he'd noted in a shout); at 12:42 p.m., after getting an A, he exuberantly left the building and hit a movie theater to celebrate with friends. Giovanni hadn't thought about that day in a long while, but now that the tool had cued him, he recalled it vividly. A year is, of course, a natural memorial moment; and if you're given an accurate cue to help reflect on a day, you're more likely to accurately re-remember it again in the future. "It's like this helps you reshape the memories of your life," he told me.

What charmed me is how such a crude signal—the mere mention of a location—could prompt so many memories: geolocation as a Proustian cookie. Again, left to our own devices, we're unlikely to bother to check year-old digital detritus. But computer code has no problem following routines. It's good at cueing memories, tickling them to recall more often and more deeply than we'd normally bother. Wegener found that people using his tool quickly formed new, creative habits around the service: They began posting more shouts—pithy, one-sentence descriptions of what they were doing—to their check-ins, since they knew that in a year, these would provide an extra bit of detail to help them remember that day. In essence, they were shouting out *to their future selves*, writing notes into a diary that would slyly present itself, one year hence, to be read.

Wegener renamed his tool Timehop and gradually added more and more forms of memories: Now it shows you pictures and status updates from a year ago, too.

Given the pattern-finding nature of computers, one can imagine increasingly sophisticated ways that our tools could automatically reconfigure and re-present our lives to us. Eric Horvitz, a Microsoft artificial intelligence researcher, has experimented with a prototype named Lifebrowser, which scours through his massive digital files to try to spot significant life events. First, you tell it which e-mails, pictures, or events in your calendar were particularly vivid; as it learns those patterns, it tries to predict what memories you'd consider to be important landmarks. Horvitz has found that "atypia"— unusual events that don't repeat—tend to be more significant, which makes sense: "No one ever needs to remember what happened at the Monday staff meeting," he jokes when I drop by his office in Seattle to see the system at work. Lifebrowser might also detect that when you've taken a lot of photos of the same thing, you were trying particularly hard to capture something important, so it'll select one representative image as important. At his desk, he shows me Lifebrowser in action. He zooms in to a single month from the previous year, and it offers up a small handful of curated events for each day: a meeting at the government's elite DARPA high-tech research department, a family visit to Whidbey Island, an e-mail from a friend announcing a surprise visit. "I would never have thought about this stuff myself, but as soon as I see it, I go, 'Oh, right—this *was* important,'" Horvitz says. The real power of digital memories will be to trigger our human ones.

In 1942, Borges published another story, about a man with perfect memory. In "Funes, the Memorious," the narrator encounters a nineteen-year-old boy who, after a horse-riding accident, discovers that he has been endowed with perfect recall. He performs astonishing acts of memory, such as reciting huge swathes of the ancient Roman text *Historia Naturalis* and describing the precise shape of a set of clouds he saw several months ago. But his immaculate memory, Funes confesses, has made him miserable. Since he's unable to forget anything, he is tortured by constantly recalling too much detail, too many minutiae, about everything. For him, forgetting would be a gift. "My memory, sir," he said, "is like a garbage heap."

Technically, the condition of being unable to forget is called hyperthymesia, and it has occasionally been found in real-life people. In the 1920s, Russian psychologist Aleksandr Luria examined Solomon Shereshevskii, a young journalist who was able to perform incredible feats of memory. Luria would present Shereshevskii with lists of numbers or words up to seventy figures long. Shereshevskii could recite the list back perfectly—not just right away, but also weeks or months later. Fifteen years after first meeting Shereshevskii, Luria met with him again. Shereshevskii sat down, closed his eyes, and accurately recalled not only the string of numbers but photographic details of the original day from years before. "You were sitting at the table and I in the rocking chair . . . You were wearing a gray suit," Shereshevskii told him. But Shereshevskii's gifts did not make him happy. Like Funes, he found the weight of so much memory oppressive. His memory didn't even make him smarter; on the contrary, reading was difficult because individual words would constantly trigger vivid memories that disrupted his attention. He "struggled to grasp" abstract concepts like *infinity* or *eternity*. Desperate to forget things, Shereshevskii would write down memories on paper and burn them, in hopes that he could destroy his past with "the magical act of burning." It didn't work.

As we begin to record more and more of our lives—intentionally and unintentionally—one can imagine a pretty bleak future. There are terrible parts of my life I'd rather not have documented (a divorce, the sudden death of my best friend at age forty); or at least, when I recall them, I might prefer my inaccurate but self-serving human memories. I can imagine daily social reality evolving into a set of weird gotchas, of the sort you normally see only on a political campaign trail. My wife and I, like many couples, bicker about who should clean the kitchen; what will life be like when there's a permanent record on tap and we can prove whose turn it is? Sure, it'd be more accurate and fair; it'd also be more picayune and crazy. These aren't idle questions, either, or even very far off. The sorts of omnipresent recording technologies that used to be experimental or figments of sci-fi are now showing up for sale on Amazon. A company named Looxcie sells a tiny camera to wear over your ear, like a Bluetooth phone mike; it buffers ten hours of video, giving the wearer an ability to rewind life like a TiVo. You can buy commercial variants of Bell's SenseCam, too.

Yet the experience of the early lifeloggers suggests that we're likely to steer a middle path with artificial memory. It turns out that even those who are rabidly trying to record everything quickly realize their psychic limits, as well as the limits of the practice's usefulness.

This is particularly true when it comes to the socially awkward aspect of lifelogging—which is that recording one's own life inevitably means recording other people's, too. Audio in particular seems to be unsettling. When Bell began his lifelogging project, his romantic partner quickly began insisting she mostly be left out of it. "We'd be talking, and she'd suddenly go, 'You didn't record that, did you?' And I'd admit, 'Yeah, I did.' 'Delete it! Delete it!'" Cathal Gurrin discovered early in his experiment that people didn't mind being on camera. "Girlfriends have been remarkably accepting of it. Some think it's really great to have their picture taken," he notes. But he gave up on trying to record audio. "One colleague of mine did it for a week, and nobody would talk to him." He laughs.

Pictures, he suspects, offer a level of plausible deniability that audio doesn't. I've noticed this, too, as a reporter. When I turn my audio recorder off during an interview, people become more open and candid, even if they're still on the record. People want their memories to be *cued*, not fully replaced; we reserve the existential pleasures of gently rewriting our history.

Gurrin argues that society will have to evolve social codes that govern artificial memory. "It's like there's now an unspoken etiquette around when you can and can't take mobile phone pictures," he suggests. Granted, these codes aren't yet very firm, and will probably never be; six years into Facebook's being a daily tool, intimate friends still disagree about whether it's fair to post drunken pictures of each other. Interestingly (or disturbingly), in our social lives we seem to be adopting concepts that used to obtain solely in institutional and legal environments. The idea of a meeting going "in camera" or "off the record" is familiar to members of city councils or corporate boards. But that language is seeping into everyday life: the popular Google Chat program added a button so users could go "off the record," for example. Viktor Mayer-Schönberger, the author of *Delete: The Virtue of Forgetting in the Digital Age*, says we'll need to engineer more artificial forgetting into our lives. He suggests that digital tools should be designed so that, when we first record something—a picture, a blog post, an instant messaging log—we're asked how long it ought to stick around: a day, a week, forever? When the time is up, it's automatically zapped into the dustbin. This way, he argues, our life traces would consist only of the stuff we've actively decided ought to stick around. It's an intriguing idea, which I will take up later when I discuss social networking and privacy.

But the truth is, research has found that people are emotional pack rats. Even when it comes to digital memories that are depressing or disturbing, they opt to preserve them. While researching his

PhD dissertation, Jason Zalinger—now a digital culture professor at the University of South Florida—got interested in Gmail, since it was the first e-mail program to actively encourage people to never delete any messages. In a sense, Gmail is the de facto lifelog for many of its users: e-mail can be quite personal, and it's a lot easier to search than photos or videos. So what, Zalinger wondered, did people do with e-mails that were emotionally fraught?

The majority kept everything. Indeed, the more disastrous a relationship, the more likely they were to keep a record—and to go back and periodically read it. One woman, Sara, had kept everything from racy e-mails traded with a married boss ("I'm talking bondage references") to e-mails from former boyfriends; she would occasionally hunt them down and reread them, as a sort of self-scrutiny. "I think I might have saved some of the painful e-mails because I wanted to show myself later, 'Wow was this guy a dick.'" The saved e-mails also, she notes, "gave me texts to analyze . . . I just read and reread until I guess I hit the point that it either stopped hurting, or I stopped looking." Another woman, Monica, explained how she'd saved all the e-mails from a partner who'd dumped her by abruptly showing up at a Starbucks with a pillowcase filled with her belongings. "I do read over those e-mails a lot," she said, "just to kind of look back, and I guess still try to figure what exactly went wrong. I won't ever get an answer, but it's nice to have tangible proof that something did happen and made an impact on my life, you know? In the beginning it was painful to read, but now it's kind of like a memory, you know?"

One man that Zalinger interviewed, Winston, had gone through a divorce. Afterward, he was torn about what to do with the e-mails from his ex-wife. He didn't necessarily want to look at them again; most divorced people, after all, want their organic memory to fade and soften the story. But he also figured, who knows? He *might* want to look at them someday, if he's trying to remember a detail or make sense of his life. In fact, when Winston thought about it, he realized there were a lot of other e-mails from his life that fit into this odd category—stuff you don't want to look at but don't want to lose, either. So he took all these emotionally difficult messages and archived them in Gmail using an evocative label: "Forget." Out of sight, out of mind, but retrievable.

It's a beautiful metaphor for the odd paradoxes and trade-offs we'll live with in a world of infinite memory. Our ancestors learned how to remember; we'll learn how to forget.

Public Thinking_

In 2003, Kenyan-born Ory Okolloh was a young law student who was studying in the United States but still obsessed with Kenyan politics. There was plenty to obsess over. Kenya was a cesspool of government corruption, ranking near the dismal bottom on the Corruption Perceptions Index. Okolloh spent hours and hours talking to her colleagues about it, until eventually one suggested the obvious: *Why don't you start a blog?*

Outside of essays for class, she'd never written anything for an audience. But she was game, so she set up a blog and faced the keyboard.

"I had zero ideas about what to say," she recalls.

This turned out to be wrong. Over the next seven years, Okolloh revealed a witty, passionate voice, keyed perfectly to online conversation. She wrote a steady stream of posts about the battle against Kenyan corruption, linking to reports of bureaucrats spending enormous sums on luxury vehicles and analyzing the "Anglo-leasing scandal," in which the government paid hundreds of millions for services—like producing a new passport system for the country—that were never delivered. When she moved back to Kenya in 2006, she began posting snapshots of such things as the bathtub-sized muddy potholes on the road to the airport. ("And our economy is supposed to be growing how exactly?") Okolloh also wrote about daily life, posting pictures of her baby and discussing the joys of living in Nairobi, including cabdrivers so friendly they'd run errands for her. She gloated nakedly when the Pittsburgh Steelers, her favorite football team, won a game.

After a few years, she'd built a devoted readership, including many Kenyans living in and out of the country. In the comments, they'd joke about childhood memories like the "packed lunch trauma" of low-income kids being sent to school with ghastly leftovers. Then in 2007, the ruling party rigged the national election and the country exploded in violence. Okolloh wrote anguished posts, incorporating as much hard information as she could get. The president imposed a media blackout, so the country's patchy Internet service was now a crucial route for news. Her blog quickly became a clearinghouse for information on the crisis, as Okolloh posted into the evening hours after coming home from work.

"I became very disciplined," she tells me. "Knowing I had these people reading me, I was very self-conscious to build my arguments, back up what I wanted to say. It was very interesting; I got this sense of obligation."

Publishers took notice of her work and approached Okolloh to write a book about her life. She turned them down. The idea terrified her. A whole book? "I have a very introverted real personality," she adds.

Then one day a documentary team showed up to interview Okolloh for a film they were producing about female bloggers. They'd printed up all her blog posts on paper. When they handed her the stack of posts, it was the size of two telephone books.

"It was huge! Humongous!" She laughs. "And I was like, *oh my*. That was the first time I had a sense of the volume of it." Okolloh didn't want to write a book, but in a sense, she already had.

The Internet has produced a foaming Niagara of writing. Consider these current rough estimates: Each day, we compose 154 billion e-mails, more than 500 million tweets on Twitter, and over 1 million blog posts and 1.3 million blog comments on WordPress alone. On Facebook, we write about 16 billion words per day. That's just in the United States: in China, it's 100 million updates each day on Sina Weibo, the country's most popular microblogging tool, and millions more on social networks in other languages worldwide, including Russia's VK. Text messages are terse, but globally they're our most frequent piece of writing: 12 billion per day.

How much writing is that, precisely? Well, doing an extraordinarily crude back-of-the-napkin calculation, and sticking only to e-mail and utterances in social media, I calculate that we're composing at least 3.6 trillion words daily, or the equivalent of 36 million books every day. The entire U.S. Library of Congress, by comparison, holds around about 35 million books.

I'm not including dozens of other genres of online composition, each of which comprises entire subgalaxies of writing, because I've never been able to find a good estimate of their size. But the numbers are equally massive. There's the world of fan fiction, the subculture in which fans write stories based on their favorite TV shows, novels, manga comics, or just about anything with a good story world and cast of characters. When I recently visited Fanfiction.net, a large repository of such writing, I calculated—again, using some equally crude napkin estimates—that there were about 325 million words' worth of stories written about the popular young-adult novel *The Hunger Games*, with each story averaging around fourteen thousand words. That's just for one book: there are thousands of other forums crammed full of writing, ranging from twenty-six thousand *Star Wars* stories to more than seventeen hundred pieces riffing off Shakespeare's works. And on top of fan fiction, there are also all the discussion boards, talmudically winding comment threads on blogs and newspapers, sprawling wikis, meticulously reported recaps of TV shows, or blow-by-blow walk-through dissections of video games; some of the ones I've used weigh in at around forty thousand words. I would hazard we're into the trillions now.

Is any of this writing good? Well, that depends on your standards, of course. I personally enjoyed Okolloh's blog and am regularly astonished by the quality and length of expression I find online, the majority of which is done by amateurs in their spare time. But certainly, measured against the prose of an Austen, Orwell, or Tolstoy, the majority of online publishing pales. This isn't surprising. The science fiction writer Theodore Sturgeon famously said something like, "Ninety percent of everything is crap," a formulation that geeks now refer to as Sturgeon's Law. Anyone who's spent time slogging through the swamp of books, journalism, TV, and movies knows that Sturgeon's Law holds pretty well even for edited and curated culture. So a global eruption of unedited, everyday self-expression is probably even more likely to produce this 90-10 split—an ocean of dreck, dotted sporadically by islands of genius. Nor is the volume of production uniform. Surveys of commenting and posting generally find that a minority of people are doing most of the creation we see online. They're ferociously overproductive, while the rest of the online crowd is quieter. Still, even given those parameters and limitations, the sheer profusion of thoughtful material that is produced every day online is enormous.

And what makes this explosion truly remarkable is what came before: comparatively little. For many people, almost nothing.

Before the Internet came along, most people rarely wrote anything at all for pleasure or intellectual satisfaction after graduating from high school or college. This is something that's particularly hard to grasp for professionals whose jobs require incessant writing, like academics, journalists, lawyers, or marketers. For them, the act of writing and hashing out your ideas seems commonplace. But until the late 1990s, this simply wasn't true of the average nonliterary person. The one exception was the white-collar workplace, where jobs in the twentieth century increasingly required more memo and report writing. But personal expression outside the workplace—in the curious genres and epic volume we now see routinely online—was exceedingly rare. For the average person there were few vehicles for publication.

What about the glorious age of letter writing? The reality doesn't match our fond nostalgia for it. Research suggests that even in the United Kingdom's peak letter-writing years—the late nineteenth century, before the telephone became common—the average citizen received barely one letter every two weeks, and that's even if we generously include a lot of distinctly unliterary business missives of the "hey, you owe us money" type. (Even the ultraliterate elites weren't pouring out epistles. They received on average two letters per week.) In the United States, the writing of letters greatly expanded after 1845, when the postal service began slashing its rates on personal letters and an increasingly mobile population needed to communicate across distances. Cheap mail was a powerful new mode of expression—though as with online writing, it was unevenly distributed, with probably only a minority of the public taking part fully, including some city dwellers who'd write and receive mail every day. But taken in aggregate, the amount of writing was remarkably small by today's standards. As the historian David Henkin notes in *The Postal Age*, the per capita volume of letters in the United States in 1860 was only 5.15 per year. "That was a huge change at the time—it was important," Henkin tells me. "But today it's the exceptional person who doesn't write five messages a day. I think a hundred years from now scholars will be swimming in a bewildering excess of life writing."

As an example of the pre-Internet age, consider my mother. She's seventy-seven years old and extremely well read—she received a terrific education in the Canadian high school system and voraciously reads novels and magazines. But she doesn't use the Internet to express herself; she doesn't write e-mail, comment on discussion threads or Facebook, post status updates, or answer questions online. So I asked her how often in the last year she'd written something of at least a paragraph in length. She laughed. "Oh, never!" she said. "I sign my name on checks or make lists—that's about it." Well, how about in the last ten years? Nothing to speak of, she recalled. I got desperate: How about twenty or thirty years back? Surely you wrote letters to family members? Sure, she said. But only about "three or four a year." In her job at a rehabilitation hospital, she jotted down the occasional short note about a patient. You could probably take all the prose she's generated since she left high school in 1952 and fit it in a single file folder.

Literacy in North America has historically been focused on reading, not writing; consumption, not production. Deborah Brandt, a scholar who researched American literacy in the 1980s and '90s, has pointed out a curious aspect of parenting: while many parents worked hard to ensure their children were regular readers, they rarely pushed them to become regular writers. You can understand the parents' point of view. In the industrial age, if you happened to write something, you were extremely unlikely to publish it. Reading, on the other hand, was a daily act crucial for navigating the world. Reading is also understood to have a moral dimension; it's supposed to make you a better person. In contrast, Brandt notes, writing was something you did mostly for work, serving an industrial purpose and not personal passions. Certainly, the people Brandt studied often enjoyed their work writing and took pride in doing it well. But without the impetus of the job, they wouldn't be doing it at all. Outside of the office, there were fewer reasons or occasions to do so.

The advent of digital communications, Brandt argues, has upended that notion. We are now a global culture of avid writers. Some of this boom has been at the workplace; the clogged e-mail inboxes of white-collar workers testifies to how much for-profit verbiage we crank out. But in our own time, we're also writing a stunning amount of material about things we're simply interested in—our hobbies, our friends, weird things we've read or seen online, sports, current events, last night's episode of our favorite TV show. As Brandt notes, reading and writing have become blended: "People read in order to generate writing; we read from the posture of the writer; we write to other people who write." Or as Francesca Coppa, a professor who studies the enormous fan fiction community, explains to me, "It's like the Bloomsbury Group in the early twentieth century, where everybody is a writer and everybody is an audience. They were all writers who were reading each other's stuff, and then writing about that, too."

We know that reading changes the way we think. Among other things, it helps us formulate thoughts that are more abstract, categorical, and logical.

So how is all this *writing* changing our cognitive behavior?

For one, it can help clarify our thinking.

Professional writers have long described the way that the act of writing forces them to distill their vague notions into clear ideas. By putting half-formed thoughts on the page, we externalize them and are able to evaluate them much more objectively. This is why writers often find that it's only when they start writing that they figure out what they want to say.

Poets famously report this sensation. "I do not sit down at my desk to put into verse something that is already clear in my mind," Cecil Day-Lewis wrote of his poetic compositions. "If it were clear in my mind, I should have no incentive or need to write about it. . . . We do not write in order to be understood; we write in order to understand." William Butler Yeats originally intended "Leda and the Swan" to be an explicitly political poem about the impact of Hobbesian individualism; in fact, it was commissioned by the editor of a political magazine. But as Yeats played around on the page, he became obsessed with the existential dimensions of the Greek myth of Leda—and the poem transformed into a spellbinding meditation on the terrifying feeling of being swept along in forces beyond your control. "As I wrote," Yeats later recalled, "bird and lady took such possession of the scene that all politics went out of it." This phenomenon isn't limited to poetry. Even the workplace that Brandt studied—including all those memos cranked out at white-collar jobs—help clarify one's thinking, as many of Brandt's subjects told her. "It crystallizes you," one said. "It crystallizes your thought."

The explosion of online writing has a second aspect that is even more important than the first, though: it's almost always done for an *audience*. When you write something online—whether it's a one-sentence status update, a comment on someone's photo, or a thousand-word post—you're doing it with the expectation that someone might read it, even if you're doing it anonymously.

Audiences clarify the mind even more. Bloggers frequently tell me that they'll get an idea for a blog post and sit down at the keyboard in a state of excitement, ready to pour their words forth. But pretty soon they think about the fact that someone's going to *read* this as soon as it's posted. And suddenly all the weak points in their argument, their clichés and lazy, autofill thinking, become painfully obvious. Gabriel Weinberg, the founder of DuckDuckGo—an upstart search engine devoted to protecting its users' privacy—writes about search-engine politics, and he once described the process neatly:

> Blogging forces you to write down your arguments and assumptions. This is the single biggest reason to do it, and I think it alone makes it worth it. You have a lot of opinions. I'm sure some of them you hold strongly. Pick

one and write it up in a post—I'm sure your opinion will change somewhat, or at least become more nuanced. When you move from your head to "paper," a lot of the hand-waveyness goes away and you are left to really defend your position to yourself.

"Hand waving" is a lovely bit of geek coinage. It stands for the moment when you try to show off to someone else a cool new gadget or piece of software you created, which suddenly won't work. Maybe you weren't careful enough in your wiring; maybe you didn't calibrate some sensor correctly. Either way, your invention sits there broken and useless, and the audience stands there staring. In a panic, you try to *describe* how the gadget works, and you start waving your hands to illustrate it: hand waving. But nobody's ever convinced. Hand waving means you've failed. At MIT's Media Lab, the students are required to show off their new projects on Demo Day, with an audience of interested spectators and corporate sponsors. For years the unofficial credo was "demo or die": if your project didn't work as intended, you died (much as stand-up comedians "die" on stage when their act bombs). I've attended a few of these events and watched as some poor student's telepresence robot freezes up and crashes . . . and the student's desperate, white-faced hand waving begins.

When you walk around meditating on an idea quietly to yourself, you do a lot of hand waving. It's easy to win an argument inside your head. But when you face a *real* audience, as Weinberg points out, the hand waving has to end. One evening last spring he rented the movie *Moneyball*, watching it with his wife after his two toddlers were in bed. He's a programmer, so the movie—about how a renegade baseball coach picked powerful players by carefully analyzing their statistics—inspired five or six ideas he wanted to blog about the next day. But as usual, those ideas were rather fuzzy, and it wasn't until he sat down at the keyboard that he realized he wasn't quite sure what he was trying to say. He was hand waving.

"Even if I was publishing it to no one, it's just the *threat* of an audience," Weinberg tells me. "If someone could come across it under my name, I have to take it more seriously." Crucially, he didn't want to bore anyone. Indeed, one of the unspoken cardinal rules of online expression is *be more interesting*—the sort of social pressure toward wit and engagement that propelled coffeehouse conversations in Europe in the nineteenth century. As he pecked away at the keyboard, trying out different ideas, Weinberg slowly realized what interested him most about the movie. It wasn't any particularly clever bit of math the baseball coach had performed. No, it was how the coach's focus on numbers had created a new way to excel at baseball. The baseball coach's behavior reminded him of how small entrepreneurs succeed: they figure out something that huge, intergalactic companies simply can't spot, because they're stuck in their old mind-set. Weinberg's process of crafting his idea—and trying to make it clever for his readers—had uncovered its true dimensions. Reenergized, he dashed off the blog entry in a half hour.

Social scientists call this the "audience effect"—the shift in our performance when we know people are watching. It isn't always positive. In live, face-to-face situations, like sports or live music, the audience effect often makes runners or musicians perform better, but it can sometimes psych them out and make them choke, too. Even among writers I know, there's a heated divide over whether thinking about your audience is fatal to creativity. (Some of this comes down to temperament and genre, obviously: Oscar Wilde was a brilliant writer and thinker who spent his life swanning about in society, drawing the energy and making the observations that made his plays and essays crackle with life; Emily Dickinson was a brilliant writer and thinker who spent her life sitting at home alone, quivering neurasthenically.)

But studies have found that particularly when it comes to analytic or critical thought, the effort of communicating to someone else forces you to think more precisely, make deeper connections, and learn more.

You can see this audience effect even in small children. In one of my favorite experiments, a group of Vanderbilt University professors in 2008 published a study in which several dozen four- and five-year-olds were shown patterns of colored bugs and asked to predict which would be next in the sequence. In one group, the children simply solved the puzzles quietly by themselves. In a second group, they were asked to explain into a tape recorder how they were solving each puzzle, a recording they could keep for themselves. And in the third group, the kids had an audience: they had to explain their reasoning to their mothers, who sat near them, listening but not offering any help. Then each group was given patterns that were more complicated and harder to predict.

The results? The children who solved the puzzles silently did worst of all. The ones who talked into a tape recorder did better— the mere act of articulating their thinking process aloud helped them think more critically and identify the patterns more clearly. But the ones who were talking to a meaningful audience—Mom— did best of all. When presented with the more complicated puzzles, on average they solved more than the kids who'd talked to themselves and about twice as many as the ones who'd worked silently.

Researchers have found similar effects with older students and adults. When asked to write for a real audience of students in another country, students write essays that are substantially longer and have better organization and content than when they're writing for their teacher. When asked to contribute to a wiki—a space that's highly public and where the audience can respond by deleting or changing your words—college students snap to attention, writing more formally and including more sources to back up their work. Brenna Clarke Gray, a professor at Douglas College in British Columbia, assigned her English students to create Wikipedia entries on Canadian writers, to see if it would get them to take the assignment more seriously. She was stunned how well it worked. "Often they're handing in these short essays without any citations, but with Wikipedia they suddenly were staying up to two a.m. honing and rewriting the entries and carefully sourcing everything," she tells me. The reason, the students explained to her, was that their audience—the Wikipedia community—was quite gimlet eyed and critical. They were harder "graders" than Gray herself. When the students first tried inputting badly sourced articles, the Wikipedians simply deleted them. So the students were forced to go back, work harder, find better evidence, and write more persuasively. "It was like night and day," Gray adds.

Sir Francis Bacon figured this out four centuries ago, quipping that "reading maketh a full man, conference a ready man, and writing an exact man."

Interestingly, the audience effect doesn't necessarily require a big audience to kick in. This is particularly true online. Weinberg, the DuckDuckGo blogger, has about two thousand people a day looking at his blog posts; a particularly lively response thread might only be a dozen comments long. It's not a massive crowd, but from his perspective it's transformative. In fact, many people have told me they feel the audience effect kick in with even a tiny handful of viewers. I'd argue that the cognitive shift in going from an audience of zero (talking to yourself) to an audience of ten people (a few friends or random strangers checking out your online post) is so big that it's actually huger than going from ten people to a million people.

This is something that the traditional thinkers of the industrial age—particularly print and broadcast journalists—have trouble grasping. For them, an audience doesn't mean anything unless it's massive. If you're writing specifically to make money, you need a large audience. An audience of ten is meaningless. Economically, it means you've failed. This is part of the thinking that causes traditional media executives to scoff at the spectacle of the "guy sitting in his living room in his pajamas writing what he thinks." But for the rest of the people in the world, who never did much nonwork writing in the first place—and who almost never did it for an audience—even a handful of readers can have a vertiginous, catalytic impact.

Writing about things has other salutary cognitive effects. For one, it improves your memory: write about something and you'll remember it better, in what's known as the "generation effect." Early evidence came in 1978, when two psychologists tested people to see how well they remembered words that they'd written down compared to words they'd merely read. Writing won out. The people who wrote words remembered them better than those who'd only read them—probably because generating text yourself "requires more cognitive effort than does reading, and effort increases memorability," as the researchers wrote. College students have harnessed this effect for decades as a study technique: if you force yourself to jot down what you know, you're better able to retain the material.

This sudden emergence of audiences is significant enough in Western countries, where liberal democracies guarantee the right to free speech. But in countries where there's less of a tradition of free speech, the emergence of networked audiences may have an even more head-snapping effect. When I first visited China to meet some of the country's young bloggers, I'd naively expected that most of them would talk about the giddy potential of arguing about human rights and free speech online. I'd figured that for people living in an authoritarian country, the first order of business, once you had a public microphone, would be to agitate for democracy.

But many of them told me it was startling enough just to suddenly be writing, in public, about the minutiae of their everyday lives—arguing with friends (and interested strangers) about stuff like whether the movie *Titanic* was too sappy, whether the fashion in the *Super Girl* competitions was too racy, or how they were going to find jobs. "To be able to speak about what's going on, what we're watching on TV, what books we're reading, what we feel about things, that is a remarkable feeling," said a young woman who had become Internet famous for writing about her sex life. "It is completely different from what our parents experienced." These young people believed in political reform, too. But they suspected that the creation of small, everyday audiences among the emerging middle-class online community, for all the seeming triviality of its conversation, was a key part of the reform process.

Once thinking is public, connections take over. Anyone who's googled their favorite hobby, food, or political subject has immediately discovered that there's some teeming site devoted to servicing the infinitesimal fraction of the public that shares their otherwise wildly obscure obsession. (Mine: building guitar pedals, modular origami, and the 1970s anime show *Battle of the Planets*). Propelled by the hyperlink—the ability of anyone to link to anyone else—the Internet is a connection-making machine.

And making connections is a big deal in the history of thought—and its future. That's because of a curious fact: If you look at the world's biggest breakthrough ideas, they often occur simultaneously to different people.

This is known as the theory of multiples, and it was famously documented in 1922 by the sociologists William Ogburn and Dorothy Thomas. When they surveyed the history of major modern inventions and scientific discoveries, they found that almost all the big ones had been hit upon by different people, usually within a few years of each other and sometimes within a few weeks. They cataloged 148 examples: Oxygen was discovered in 1774 by Joseph Priestley in London and Carl Wilhelm Scheele in Sweden (and Scheele had hit on the idea several years earlier). In 1610 and 1611, four different astronomers—including Galileo—independently discovered sunspots. John Napier and Henry Briggs developed logarithms in Britain while Joost Bürgi did it independently in Switzerland. The law of the conservation of energy was laid claim to by four separate people in 1847. And radio was invented at the same time around 1900 by Guglielmo Marconi and Nikola Tesla.

Why would the same ideas occur to different people at the same time? Ogburn and Thomas argued that it was because our ideas are, in a crucial way, partly products of our environment. They're "inevitable." When they're ready to emerge, they do. This is because we, the folks coming up with the ideas, do not work in a sealed-off, Rodin's *Thinker* fashion. The things we think about are deeply influenced by the state of the art around us: the conversations taking place among educated folk, the shared information, tools, and technologies at hand. If four astronomers discovered sunspots at the same time, it's partly because the quality of lenses in telescopes in 1611 had matured to the point where it was finally possible to pick out small details on the sun and partly because the question of the sun's role in the universe had become newly interesting in the wake of Copernicus's heliocentric theory. If radio was developed at the same time by two people, that's because the basic principles that underpin the technology were also becoming known to disparate thinkers. Inventors knew that electricity moved through wires, that electrical currents caused fields, and that these seemed to be able to jump distances through the air. With that base of knowledge, curious minds are liable to start wondering: Could you use those signals to communicate? And as Ogburn and Thomas noted, there are a *lot* of curious minds. Even if you assume the occurrence of true genius is pretty low (they estimated that one person in one hundred was in the "upper tenth" for smarts), that's still a heck of a lot of geniuses.

When you think of it that way, what's strange is not that big ideas occurred to different people in different places. What's strange is that this didn't happen *all the time, constantly.*

But maybe it did—and the thinkers just weren't yet in contact. Thirty-nine years after Ogburn and Thomas, sociologist Robert Merton took up the question of multiples. (He's the one who actually coined the term.) Merton noted an interesting corollary, which is that when inventive people aren't aware of what others are working on, the pace of innovation slows. One survey of mathematicians, for example, found that 31 percent complained that they had needlessly duplicated work that a colleague was doing—because they weren't aware it was going on. Had they known of each other's existence, they could have collaborated and accomplished their calculations more quickly or with greater insight.

As an example, there's the tragic story of Ernest Duchesne, the original discoverer of penicillin. As legend has it, Duchesne was a student in France's military medical school in the mid-1890s when he noticed that the stable boys who tended the army's horses did something peculiar: they stored their saddles in a damp, dark room so that mold would grow on their undersurfaces. They did this, they explained, because the mold helped heal the horses' saddle sores. Duchesne was fascinated and conducted an experiment in which he treated sick guinea pigs with a solution made from mold— a rough form of what we'd now call penicillin. The guinea pigs healed completely. Duchesne wrote up his findings in a PhD thesis, but because he was unknown and young—only twenty-three at the time—the French Institut Pasteur wouldn't acknowledge it. His research vanished, and Duschesne died fifteen years later during his military service, reportedly of tuberculosis. It would take another thirty-two years for Scottish scientist Alexander Fleming to rediscover penicillin, independently and with no idea that Duchesne had already done it. Untold millions of people died in those three decades of diseases that could have been cured. Failed networks kill ideas.

When you can resolve multiples and connect people with similar obsessions, the opposite happens. People who are talking and writing and working on the same thing often find one another, trade ideas, and collaborate. Scientists have for centuries intuited the power of resolving multiples, and it's part of the reason that in the seventeenth century they began publishing scientific journals and setting standards for citing the similar work of other scientists. Scientific journals and citation were a successful attempt to create a worldwide network, a mechanism for not just thinking in public but doing so in a connected way. As the story of Duchesne shows, it works pretty well, but not all the time.

Today we have something that works in the same way, but for everyday people: the Internet, which encourages public thinking and resolves multiples on a much larger scale and at a pace more dementedly rapid. It's now the world's most powerful engine for putting heads together. Failed networks kill ideas, but successful ones trigger them.

As an example of this, consider what happened next to Ory Okolloh. During the upheaval after the rigged Kenyan election of 2007, she began tracking incidents of government violence. People called and e-mailed her tips, and she posted as many as she could. She wished she had a tool to do this automatically—to let anyone post an incident to a shared map. So she wrote about that:

> Google Earth supposedly shows in great detail where the damage is being done on the ground. It occurs to me that it will be useful to keep a record of this, if one is thinking long-term. For the reconciliation process to occur at the local level the truth of what happened will first have to come out. Guys looking to do something—any techies out there willing to do a mashup of where the violence and destruction is occurring using Google Maps?

One of the people who saw Okolloh's post was Erik Hersman, a friend and Web site developer who'd been raised in Kenya and lived in Nairobi. The instant Hersman read it, he realized he knew some-one who could make the idea a reality. He called his friend David Kobia, a Kenyan programmer who was working in Birmingham, Alabama. Much like Okolloh, Kobia was interested in connecting Kenyans to talk about the country's crisis, and he had created a dis-cussion site devoted to it. Alas, it had descended into political toxic-ity and calls for violence, so he'd shut it down, depressed by having created a vehicle for hate speech. He was driving out of town to visit some friends when he got a call from Hersman. Hersman explained Okolloh's idea—a map-based tool for reporting violence—and Ko-bia immediately knew how to make it happen. He and Hersman contacted Okolloh, Kobia began frantically coding with them, and within a few days they were done. The tool allowed anyone to pick a location on a Google Map of Kenya, note the time an incident oc-curred, and describe what happened. They called it Ushahidi—the Swahili word for "testimony."

Within days, Kenyans had input thousands of incidents of electoral violence. Soon after, Ushahidi attracted two hundred thousand dollars in nonprofit funds and the trio began refining it to accept reports via everything from SMS to Twitter. Within a few years, Ushahidi had become an indispensable tool worldwide, with governments and nonprofits relying on it to help determine where to send assistance. After a massive earthquake hit Haiti in 2010, a Ushahidi map, set up within hours, cataloged twenty-five thousand text messages and more than four million tweets over the next month. It has become what Ethan Zuckerman, head of MIT's Center for Civic Media, calls "one of the most globally significant technology projects."

The birth of Ushahidi is a perfect example of the power of public thinking and multiples. Okolloh could have simply wandered around wishing such a tool existed. Kobia could have wandered around wishing he could use his skills to help Kenya. But because Okolloh was thinking out loud, and because she had an audience of like-minded people, serendipity happened.

The tricky part of public thinking is that it works best in situations where people aren't worried about "owning" ideas. The existence of multiples—the knowledge that people out there are puzzling over the same things you are—is enormously exciting if you're trying to solve a problem or come to an epiphany. But if you're trying to make *money*? Then multiples can be a real problem. Because in that case you're trying to stake a claim to ownership, to being the first to think of something. Learning that other people have the same idea can be anything from annoying to terrifying.

Scientists themselves are hardly immune. Because they want the fame of discovery, once they learn someone else is working on a similar problem, they're as liable to *compete* as to collaborate—and they'll bicker for decades over who gets credit. The story of penicillin illustrates this as well. Three decades after Duchesne made his discovery of pencillin, Alexander Fleming in 1928 stumbled on it again, when some mold accidentally fell into a petri dish and killed off the bacteria within. But Fleming didn't seem to believe his discovery could be turned into a lifesaving medicine, so, remarkably, he never did any animal experiments and soon after dropped his research entirely. Ten years later, a pair of scientists in Britain— Ernest Chain and Howard Florey—read about Fleming's work, intuited that penicillin could be turned into a medicine, and quickly created an injectable drug that cured infected mice. After the duo published their work, Fleming panicked: someone else might get credit for his discovery! He hightailed it over to Chain and Florey's lab, greeting them with a wonderfully undercutting remark: "I have come to see what you've been doing with *my* old penicillin." The two teams eventually worked together, transforming penicillin into a mass-produced drug that saved countless lives in World War II. But for years, even after they all received a Nobel Prize, they jousted gently over who ought to get credit.

The business world is even more troubled by multiples. It's no wonder; if you're trying to make some money, it's hardly comforting to reflect on the fact that there are hundreds of others out there with precisely the same concept. Patents were designed to prevent someone else from blatantly infringing on your idea, but they also function as a response to another curious phenomenon: *unintentional* duplication. Handing a patent on an invention to one person creates artificial scarcity. It is a crude device, and patent offices have been horribly abused in recent years by "patent trolls"; they're people who get a patent for something (either by conceiving the idea themselves, or buying it) without any intention of actually producing the invention—it's purely so they can sue, or soak, people who go to market with the same concept. Patent trolls employ the concept of multiples in a perverted reverse, using the common nature of new ideas to hold all inventors hostage.

I've talked to entrepreneurs who tell me they'd like to talk openly online about what they're working on. They *want* to harness multiples. But they're worried that someone will take their idea and execute it more quickly than they can. "I know I'd get better feedback on my project if I wrote and tweeted about it," one once told me, "but I can't risk it." This isn't universally true; some start-up CEOs have begun trying to be more open, on the assumption that, as Bill Joy is famously reported quipping, "No matter who you are, most of the smartest people work for someone else." They know that talking about a problem makes it more likely you'll hook up with someone who has an answer.

But on balance, the commercial imperative to "own" an idea explains why public thinking has been a boon primarily for everyday people (or academics or nonprofits) pursuing their amateur passions. If you're worried about making a profit, multiples dilute your special position in the market; they're depressing. But if you're just trying to improve your thinking, multiples are exciting and catalytic. Everyday thinkers online are thrilled to discover someone else with the same idea as them.

We can see this in the history of "giving credit" in social media. Every time a new medium for public thinking has emerged, early users set about devising cordial, Emily Post–esque protocols. The first bloggers in the late 1990s duly linked back to the sources where they'd gotten their fodder. They did it so assiduously that the creators of blogging software quickly created an automatic "trackback" tool to help automate the process. The same thing happened on Twitter. Early users wanted to hold conversations, so they began using the @ reply to indicate they were replying to someone—and then to credit the original user when retweeting a link or pithy remark. Soon the hashtag came along—like #stupidestthingivedone today or #superbowl—to create floating, ad hoc conversations. All these innovations proved so popular that Twitter made them a formal element of its service. We so value conversation and giving credit that we hack it into any system that comes along.

Stanford University English professor Andrea Lunsford is one of America's leading researchers into how young people write. If you're worried that college students today can't write as well as in the past, her work will ease your mind. For example, she tracked down studies of how often first-year college students made grammatical errors in freshman composition essays, going back nearly a century. She found that their error rate has barely risen at all. More astonishingly, today's freshman-comp essays are over six times longer than they were back then, and also generally more complex. "Student essayists of the early twentieth century often wrote essays on set topics like 'spring flowers,'" Lunsford tells me, "while those in the 1980s most often wrote personal experience narratives. Today's students are much more likely to write essays that present an argument, often with evidence to back them up"—a much more challenging task. And as for all those benighted texting short forms, like LOL, that have supposedly metastasized in young people's formal writing? Mostly nonexistent. "Our findings do not support such fears," Lunsford wrote in a paper describing her research, adding, "In fact, we found almost no instances of IM terms." Other studies have generally backed up Lunsford's observations: one analyzed 1.5 million words from instant messages by teens and found that even there, only 3 percent of the words used were IM-style short forms. (And while spelling and capitalization could be erratic, not all was awry; for example, youth substituted "u" for "you" only 8.6 percent of the time they wrote the word.) Others have found that kids who message a lot appear to have have slightly better spelling and literacy abilities than those who don't. At worst, messaging—with its half-textual, half-verbal qualities—might be reinforcing a preexisting social trend toward people writing more casually in otherwise formal situations, like school essays or the workplace.

In 2001, Lunsford got interested in the writing her students were doing everywhere—not just in the classroom, but outside it. She began the five-year Stanford Study of Writing, and she convinced 189 students to give her copies of everything they wrote, all year long, in any format: class papers, memos, e-mails, blog and discussion-board posts, text messages, instant-message chats, and more. Five years later, she'd collected nearly fifteen thousand pieces of writing and discovered something notable: The amount of writing kids did outside the class was huge. In fact, roughly 40 percent of everything they wrote was for pleasure, leisure, or socializing. "They're writing so much more than students before them ever did," she tells me. "It's stunning."

Lunsford also finds it striking how having an audience changed the students' writing outside the classroom. Because they were often writing for other people—the folks they were e-mailing with or talking with on a discussion board—they were adept at reading the tempo of a thread, adapting their writing to people's reactions. For Lunsford, the writing strategies of today's students have a lot in common with the Greek ideal of being a smart rhetorician: knowing how to debate, to marshal evidence, to listen to others, and to concede points. Their writing was constantly in dialogue with others.

"I think we are in the midst of a literacy revolution the likes of which we have not seen since Greek civilization," Lunsford tells me. The Greek oral period was defined by knowledge that was formed face-to-face, in debate with others. Today's online writing is like a merging of that culture and the Gutenberg print one. We're doing more jousting that takes place in text but is closer in pacing to a face-to-face conversation. No sooner does someone assert something than the audience is reacting—agreeing, challenging, hysterically criticizing, flattering, or being abusive.

The upshot is that public thinking is often less about product than *process*. A newspaper runs a story, a friend posts a link on Facebook, a blogger writes a post, and it's interesting. But the real intellectual action often takes place in the comments. In the spring of 2011, a young student at Rutgers University in New Jersey was convicted of using his webcam to spy on a gay roommate, who later committed suicide. It was a controversial case and a controversial verdict, and when the *New York Times* wrote about it, it ran a comprehensive story more than 1,300 words long. But the readers' comments were many times larger—1,269 of them, many of which were remarkably nuanced, replete with complex legal and ethical arguments. I learned considerably more about the Rutgers case in a riveting half hour of reading *New York Times* readers debate the case than I learned from the article, because the article—substantial as it was—could represent only a small number of facets of a terrifically complex subject.

Socrates might be pleased. Back when he was alive, twenty-five hundred years ago, society had begun shifting gradually from an oral mode to a written one. For Socrates, the advent of writing was dangerous. He worried that text was too inert: once you wrote something down, that text couldn't adapt to its audience. People would read your book and think of a problem in your argument or want clarifications of your points, but they'd be out of luck. For Socrates, this was deadly to the quality of thought, because in the Greek intellectual tradition, knowledge was formed in the cut and thrust of debate. In Plato's *Phaedrus*, Socrates outlines these fears:

> I cannot help feeling, Phaedrus, that writing is unfortunately like painting; for the creations of the painter have the attitude of life, and yet if you ask them a question they preserve a solemn silence. And the same may be said of speeches. You would imagine that they had intelligence, but if you want to know anything and put a question to one of them, the speaker always gives one unvarying answer. And when they have been once written down they are tumbled about anywhere among those who may or may not understand them, and know not to whom they should reply, to whom not: and, if they are maltreated or abused, they have no parent to protect them; and they cannot protect or defend themselves.

Today's online writing meets Socrates halfway. It's print*ish*, but with a roiling culture of oral debate attached. Once something interesting or provocative is published—from a newspaper article to a book review to a tweet to a photo—the conversation begins, and goes on, often ad infinitum, and even the original authors can dive in to defend and extend their writing.

The truth is, of course, that knowledge has always been created via conversation, argument, and consensus. It's just that for the last century of industrial-age publishing, that process was mostly hidden from view. When I write a feature for a traditional print publication like *Wired* or *The New York Times*, it involves scores of conversations, conducted through e-mail and on the phone. The editors and I have to agree upon what the article will be about; as they edit the completed piece, the editors and fact-checkers will fix mistakes and we'll debate whether my paraphrase of an interviewee's point of view is too terse or glib. By the time we're done, we'll have generated a conversation about the article that's at least as long as the article itself (and probably far longer if you transcribed our phone calls). The same thing happens with every book, documentary, or scientific paper—but because we don't see the sausage being made, we in the audience often forget that most information is forged in debate. I often wish traditional publishers let their audience see the process. I suspect readers would be intrigued by how magazine fact-checkers improve my columns by challenging me on points of fact, and they'd understand more about why material gets left out of a piece—or left in it.

Wikipedia has already largely moved past its period of deep suspicion, when most academics and journalists regarded it as utterly untrustworthy. Ever since the 2005 story in *Nature* that found Wikipedia and the *Encyclopedia Britannica* to have fairly similar error rates (four errors per article versus three, respectively), many critics now grudgingly accept Wikipedia as "a great place to start your research, and the worst place to end it." Wikipedia's reliability varies heavily across the site, of course. Generally, articles with large and active communities of contributors are more accurate and complete than more marginal ones. And quality varies by subject matter; a study commissioned by the Wikipedia Foundation itself found that in the social sciences and humanities, the site is 10 to 16 percent less accurate than some expert sources.

But as the author David Weinberger points out, the deeper value of Wikipedia is that it makes transparent the arguments that go into the creation of any article: click on the "talk" page and you'll see the passionate, erudite conversations between Wikipedians as they hash out an item. Wikipedia's process, Weinberger points out, is a part of its product, arguably an indispensable part. Whereas the authority of traditional publishing relies on expertise—*trust us because our authors are vetted by our experience, their credentials, or the marketplace*—conversational media gains authority by revealing its mechanics. James Bridle, a British writer, artist, and publisher, made this point neatly when he took the entire text of every edit of

Wikipedia's much-disputed entry on the Iraq War during a five-year period and printed it as a set of twelve hardcover books. At nearly seven thousand pages, it was as long as an encyclopedia itself. The point, Bridle wrote, was to make visible just how much debate goes into the creation of a factual record: "This is historiography. This is what culture actually looks like: a process of argument, of dissenting and accreting opinion, of gradual and not always correct codification." Public thinking is messy, but so is knowledge.

I'm not suggesting here, as have some digital utopians (and dystopians), that traditional "expert" forms of thinking and publishing are obsolete, and that expertise will corrode as the howling hive mind takes over. Quite the opposite. I work in print journalism, and now in print books, because the "typographical fixity" of paper—to use Elizabeth Eisenstein's lovely phrase—is a superb tool for focusing the mind. Constraints can impose creativity and rigor. When I have only six hundred words in a magazine column to make my point, I'm forced to make decisions about what I'm willing to commit to print. Slowing down also gives you time to consult a ton of sources and intuit hopefully interesting connections among them. The sheer glacial nature of the enterprise—spending years researching a book and writing it—is a cognitive strength, a gift that industrial processes gave to civilization. It helps one escape the speed loop of the digital conversation, where it's easy to fall prey to what psychologists call recency: Whatever's happening *right now* feels like the most memorable thing, so responding right now feels even *more* urgent. (This is a problem borrowed from face-to-face conversation: You won't find a lot of half-hour-long, thoughtful pauses in coffeehouse debates either.) And while traditional "expert" media are going to evolve in form and style, I doubt they're going to vanish, contrary to some of the current hand-wringing and gloating over that prospect. Business models for traditional reportage might be foundering, but interest is not: one analysis by HP Labs looked at Twitter's "trending topics" and found that a majority of the most retweeted sources were mainstream news organizations like CNN, *The New York Times*, and Reuters.

The truth is that old and new modes of thinking aren't mutually exclusive. Knowing when to shift between public and private thinking—when to blast an idea online, when to let it slow bake—is a crucial new skill: cognitive diversity. When I get blocked while typing away at a project on my computer, I grab a pencil and paper, so I can use a tactile, swoopy, this-connects-to-that style of writing to unclog my brain. Once an idea is really flowing on paper, I often need to shift to the computer, so my seventy-words-per-minute typing and on-tap Google access can help me move swiftly before I lose my train of thought.

Artificial intelligence pioneer Marvin Minsky describes human smarts as stemming from the various ways our brains will tackle a problem; we'll simultaneously throw logic, emotion, metaphor, and crazy associative thinking at it. This works with artificial thinking tools, too. Spent too much time babbling online? Go find a quiet corner and read. Spent a ton of time working quietly alone? Go bang your ideas against other people online.

Ethan Hein is a musician who lives not far from me in Brooklyn. He teaches music and produces songs and soundtracks for indie movies and off-Broadway shows.

But most people know him as a guy who answers questions.

Tons of them. From strangers.

Hein is an enthusiastic poster on Quora, one of the current crop of question-answering sites: anyone can show up and ask a question, and anyone can answer. Hein had long been an online extrovert, blogging about music and tweeting. But he could also be, like many of us, lazy about writing. "I was always a half-assed journal keeper," he tells me. "It was like, I should write something—wait a minute, what's on TV?" But in early 2011 he stumbled upon Quora and found the questions perversely stimulating. (Question: "What does the human brain find exciting about syncopated rhythm and breakbeats?" Hein's answer began: "Predictable unpredictability. The brain is a pattern-recognition machine . . .") Other times, he chimed in on everything from neuroscience and atheism to "What is it like to sleep in the middle of a forest?" (A: "Sleeping in the woods gratifies our biophilia.") Within a year, he was hooked.

"I will happily shuffle through the unanswered questions as a form of *entertainment*," Hein says. "My wife is kind of worried about me. But I'm like, 'Look, I'd be using this time to play *World of Warcraft*. And this is better—this is contributing. To the world!'" He even found that answering questions on Quora invigorated his blogging, because once he'd researched a question and pounded out a few paragraphs, he could use the answer as the seed for a new post. In barely one year he'd answered over twelve hundred questions and written about ninety thousand words. I tell him that's the length of a good-sized nonfiction hardcover book, and, as with Ory Okolloh and her two telephone books' worth of online writing, he seems stunned.

Public thinking is powerful, but it's hard to do. It's *work*. Sure, you get the good—catalyzing multiples, learning from the feedback. But it can be exhausting. Digital tools aren't magical pixie dust that makes you smarter. The opposite is true: they give up the rewards only if you work hard and master them, just like the cognitive tools of previous generations.

But as it turns out, there are structures that can make public thinking easier—and even irresistible.

Question answering is a powerful example. In the 1990s, question-answering sites like Answerbag.com began to emerge; by now there are scores of them. The sheer volume of questions answered is remarkable: over one billion questions have been answered at the English version of Yahoo Answers, with one study finding the average answerer has written about fifty-one replies. In Korea, the search engine Naver set up shop in 1999 but realized there weren't very many Korean-language Web sites in existence, so it set up a question-answering forum, which became one of its core offerings. (And since all those questions are hosted in a proprietary database that Google can't access, Naver has effectively sealed Google out from the country, a neat trick.) Not all the answers, or questions, are good; Yahoo Answers in particular has become the butt of jokes for hosting spectacularly illiterate queries ("I CAN SMELL EVERYTHING MASSIVE HEAD ACHE?") or math students posting homework questions, hoping they'll be answered. (They usually are.) But some, like Quora, are known for cultivating thought-provoking questions and well-written answers. One of my favorite questions was "Who is history's greatest badass, and why?"—which provoked a twenty-two-thousand-word rush of answers, one of which described former U.S. president Theodore Roosevelt being shot by a would-be assassin before a speech and then, bleeding profusely, continuing to give the 1.5-hour-long address.

Why do question sites produce such outpourings of answers? It's because the format is a clever way of encouraging people to formalize and share knowledge. People walk around with tons of information and wisdom in their heads but with few outlets to show it off. Having your own Web site is powerful, but comparatively few people are willing to do the work. They face the blank-page problem. What should I say? Who *cares* what I say? In contrast, when you see someone asking a question on a subject you know about, it catalyzes your desire to speak up.

"Questions are a really useful service for curing writer's block," as Charlie Cheever, the soft-spoken cofounder of Quora, tells me. "You might think you want to start a blog, but you wind up being afraid to write a blog post because there's this sense of, who asked *you?*" Question answering provides a built-in, instant audience of at least one—the original asker. This is another legacy of Plato's Socratic dialogues, in which Socrates asks questions of his debating partners (often faux-naive, concern-trolling ones, of course) and they pose questions of him in turn. Web authors long ago turned this into a literary form that has blossomed: the FAQ, a set of mock-Socratic questions authors pose to themselves as a way of organizing information.

It's an addictive habit, apparently. Academic research into question-answering sites has found that answering begets answering: people who respond to questions are likely to stick around for months and answer even more. Many question-answering sites have a psychological architecture of rewards, such as the ability of members to give positive votes (or award "points") for good answers. But these incentives may be secondary to people's altruism and the sheer joy of helping people out, as one interview survey of Naver users discovered. The Naver users said that once they stumbled across a question that catalyzed their expertise, they were hooked; they couldn't help responding. "Since I was a doctor, I was browsing the medical directories. I found a lot of wrong answers and information and was afraid they would cause problems," as one Naver contributor said. "So I thought I'd contribute in fixing it, hoping that it'd be good for the society." Others found that the act of writing answers helped organize their own thoughts—the generation effect in a nutshell. "My first intention [in answering] was to organize and review my knowledge and practice it by explaining it to others," one explained.

These sites have formalized question answering as a vehicle for public thinking, but they didn't invent it. In almost any online community, answering questions frequently forms the backbone of conversation, evolving on a grassroots level. Several years ago while reading YouBeMom, an anonymous forum for mothers, I noticed that users had created a clever inversion of the question-answering format: a user would post a description of their job and ask if anyone had questions. The ploy worked in both directions, encouraging people to ask questions they might never have had the opportunity to ask. The post "ER nurse here—questions?" turned into a sprawling discussion, hundreds of postings long, about the nurse's bloodiest accidents, why gunshot attacks were decreasing, and how ballooning ER costs are destroying hospital budgets. (An even more spellbinding conversation emerged the night a former prostitute opened up the floor for questions.) Though it's hard to say where it emerged, the "I am a . . ." format has become, like the FAQ, another literary genre the Internet has ushered into being; on the massive discussion board Reddit, there are dozens of "IAmA" threads started each day by everyone from the famous (the comedian Louis C.K., Barack Obama) to people with intriguing experiences ("IAmA Female Vietnam Veteran"; "IAmA former meth lab operator"; "IAmA close friend of Charlie Sheen since 1985").

I'm focusing on question answering, but what's really at work here is what publisher and technology thinker Tim O'Reilly calls the "architecture of participation." The future of public thinking hinges on our ability to create tools that bring out our best: that encourage us to organize our thoughts, create audiences, make connections. Different forms encourage different styles of talk.

Microblogging created a torrent of public thinking by making a virtue of its limits. By allowing people to write only 140 characters at a time, Twitter neatly routed around the "blank page" problem: everybody can think of at least *that* many words to say. Facebook provoked a flood of writing by giving users audiences composed of people they already knew well from the offline world, people they knew cared about what they had to say. Texting offered a style of conversation that was more convenient than voice calls (and cheaper, in developing countries), and the asynchronicity created pauses useful for gathering your thoughts (or waiting until your boss's back was turned so you could sneak in a conversation). One size doesn't fit all, cognitively speaking. I know people who engage in arguments about music or politics with friends on Facebook because it's an extension of offline contact, while others find the presence of friends claustrophobic; they find it more freeing and stimulating to talk with comparative strangers on open-ended discussion boards.

Clearly, public speech can be enormously valuable. But what about the stuff that isn't? What about the *repellent* public speech? When you give everyday people the ability to communicate, you release not just brilliant bons mots and incisive conversations, but also ad hominem attacks, fury, and "trolls"—people who jump into discussion threads solely to destabilize them. The combination of distance and pseudonymity (or sometimes total anonymity) can unlock people's worst behavior, giving them license to say brutal things they'd never say to someone's face.

This abuse isn't evenly distributed. It's much less often directed at men, particularly white men like me. In contrast, many women I know—probably most—find that being public online inevitably attracts a wave of comments, ranging from dismissal to assessments of their appearance to flat-out rape threats. This is particularly true if they're talking about anything controversial or political. Or even intellectual: "An opinion, it seems, is the short skirt of the Internet," as Laurie Penny, a British political writer, puts it. This abuse is also heaped on blacks and other minorities in the United States, or any subordinated group. Even across lines of party politics, discussion threads quickly turn toxic in highly personal ways.

How do we end this type of abuse? Alas, we probably can't, at least not completely—after all, this venom is rooted in real-world

biases that go back centuries. The Internet didn't create these prejudices; it gave them a new stage.

But there are, it turns out, techniques to curtail online abuse,
sometimes dramatically. In fact, some innovators are divining,
through long experience and experimentation, key ways of managing conversation online—not only keeping it from going septic, but
improving it.

Consider the example of Ta-Nahesi Coates. Coates is a senior editor at *The Atlantic Monthly*, a magazine of politics and culture; he
ran a personal blog for years and moved it over to the *Atlantic* five
years ago. Coates posts daily on a dizzying array of subjects: movies, politics, economic disparities, the Civil War, TV shows, favorite
snippets of poetry, or whether pro football is too dangerous to play.
Coates, who is African American, is also well known as an eloquent
and incisive writer on race, and he posts about that frequently. Yet
his forum is amazingly abuse-free: comments spill into the hundreds
without going off the rails. "This is the most hot-button issue in
America, and folks have managed to keep a fairly level head," he
tells me.

The secret is the work Coates puts into his discussion board.
Before he was a blogger himself, he'd noticed the terrible comments
at his favorite political blogs, like that of Matt Yglesias. "Matt
could be talking about parking and urban issues, and he'd have ten
comments, and somebody would invariably say something racist."
Coates realized that negative comments create a loop: they poison
the atmosphere, chasing off productive posters.

So when he started his own personal blog, he decided to break
that loop. The instant he saw something abusive, he'd delete it, banning repeat offenders. Meanwhile, he went out of his way to encourage the smart folks, responding to them personally and publicly,
so they'd be encouraged to stay and talk. And Coates was unfail-

ingly polite and civil himself, to help set community standards. Soon several dozen regular commenters emerged, and they got to know each other, talking as much to each other as to Coates. (They've even formed their own Facebook group and have held "meet-ups.") Their cohesion helped cement the culture of civility even more; any troll today who looks at the threads can quickly tell this community isn't going to tolerate nastiness. The *Atlantic* also deploys software that lets users give an "up" vote to the best comments, which further helps reinforce quality. Given that the community has good standards, the first comment thread you'll see at the bottom of a Coates post is likely to be the cleverest—and not, as at sites that don't manage their comments and run things chronologically, the first or last troll to have stopped by.

This is not to say it's a love fest or devoid of conflict. The crowd argues heatedly and often takes Coates to task for his thinking; he cites their feedback in his own posts. "Being a writer does not mean you are smarter than everyone else. I *learn* things from these people," he notes. But the debate transpires civilly and without name-calling. These days, Coates still tends the comments and monitors them but rarely needs to ban anyone. "It's much easier," he adds.

What exactly do you call what Coates is doing, this mix of persuasion, listening, and good hosting, like someone skillfully tending bar? A few years ago, three Internet writers and thinkers—Deb Schultz, Heather Gold, and Kevin Mark—brainstormed on what to label it. On the suggestion of Theresa Nielsen Hayden, a longtime host of online communities, they settled on a clever term: "tummeling," derived from the Yiddish *tummler*, the person at a party responsible for keeping the crowd engaged and getting them dancing at a wedding. Tummlers are the social adepts of online conversation. "They're catalysts and bridge builders," Schultz tells me. "It's not about technology. It's about the human factor." They know how to be empathetic, how to draw people out: "A good tummler reads the room," Gold adds. "Quieter people have a disproportionately strong impact on conversational flow when drawn out and heard."

Look behind any high-functioning discussion forum online and you'll find someone doing tummeling. Without it, you get chaos. That's why YouTube is a comment cesspool; there is no culture of moderating comments. It's why you frequently see newspaper Web pages filled with toxic comments. They haven't assigned anyone to be the tummler.

Newspapers and YouTube also have another problem, which is that they're always trying to get bigger. But as Coates and others have found, conversation works best when it's smaller. Only in a more tightly knit group can participants know each other. Newspapers, in contrast, work under the advertising logic of "more is better." This produces unfocused, ad hoc, drive-by audiences that can never be corralled into community standards. Coates jokes about going to a major U.S. newspaper and seeing a link to the discussion threads—*Come on in! We have 2,000 comments!* "That's a bar I don't want to go into! They don't have any security!" he says. These sites are trying for scale—but conversation doesn't scale.

There are other tools emerging to help manage threads, such as requiring real name identity, as with Facebook comments; removing anonymity can bring in accountability, since people are less likely to be abusive if their actual name is attached to the abuse. Mind you, Coates isn't opposed per se to anonymity or to crazy, free-range places like Reddit. "Those environments catalyze a lot of rancor, sure, but also candor. The fact that places like that exist might make it even easier to do what I do," he notes.

Tummeling isn't a total solution. It works only when you control the space and can kick out undesirables. Services like Twitter are more open and thus less manageable. But even in those spaces, tummeling is a digital-age skill that we will increasingly need to learn, even formally *teach*; if this aspect of modern civics became widespread enough, it could help reform more and more public spaces online. There's a pessimistic view, too. You could argue that the first two decades of open speech have set dreadful global standards and that the downsides of requiring targeted groups—say, young women—to navigate so much hate online aren't worth the upsides of public speech. That's a reasonable caveat. When it comes to public thinking, you need to accept the bad with the good, but there's a lot of bad to accept.

What tools will create new forms of public thinking in the years to come? With mobile phones, our personal geography is becoming newly relevant in a new way. GPS turns your location into a fresh source of multiples, because it can figure out if there are other people nearby sharing your experience (say, at a concert or a park). An early success of this kind was Grindr, a phone app that lets gay men broadcast their location and status messages and locate other gay men nearby (proving again the technology truism that sex and pornography are always at the forefront of tech innovation).

The ability of phones to broadcast their location has even weirder effects, because it can turn geography into a message board, with apps that embed conversations in specific physical spaces. For example, when the Occupy Wall Street movement flared in New York City, some of the activists began using a mobile app called Vibe that let them post anonymous messages that were tagged to physical locations around Wall Street: they'd discuss where police were about to crack down or leave notes describing events they'd seen. This is bleeding into everyday life, with services that let people embed photos and thoughts on maps and engage in location-based conversations. It's the first stage of conversational "augmented reality": public thinking woven into our real-world public space.

I also suspect that as more forms of media become digital, they'll become sites for public thinking—particularly digital books. Books have always propelled smart conversations; the historic, face-to-face book club has migrated rapidly online, joining the sprawling comments at sites like Goodreads. But the pages of e-books are themselves likely to become the sites of conversations. Already readers of many e-books—on the Kindle, the Nook, and other e-readers—share comments and highlights. Marginalia may become a new type of public thinking, with the smartest remarks from other readers becoming part of how we make sense of a book. (Bob Stein, head of the Institute for the Future of the Book, imagines a cadre of marginaliasts becoming so well liked that people pay to read their markups.) The truth is, whatever new digital tools come around, curious people are going to colonize them. We're social creatures, so we think socially.

But there's one interesting kink. For most of this chapter I've been talking about one type of publishing—writing in text. It's one of our oldest and most robust tools for recording and manipulating ideas. But the digital age is also producing a Cambrian explosion in different media that we're using to talk, and think, with each other—including images, video, and data visualization. The difference is, while we're taught in school how to write and read, our traditional literacy focuses less on these new modes of publishing. We're working them out on our own for now and discovering just how powerful they can be.

The New Literacies_

How do you tackle a problem that affects the fabric of democracy but also happens to be, well, boring?

Ask Costas Panagopoulos. A professor of political science at Fordham University in New York, Panagopolous is an expert on gerrymandering, the tawdry two-hundred-year-old political phenomenon by which politicians redraw the boundaries of their districts in order to exclude anyone who won't vote for them. In theory, redistricting isn't harmful; indeed, laws require the regular rejiggering of maps to make sure that as the population shifts, it's adequately represented. But in practice, politicians manipulate this process in order to cement their own power. In the United States, Democrats try to herd liberal urbanites and blacks into their districts' boundaries while pushing out gun-loving rural folk. Republicans do the reverse.

Politicians worldwide love this trick, but in New York State they've made it an art form. In the last fifty years, they've redrawn their electoral districts in such nakedly self-serving ways it's a statewide joke. (One district was redrawn so tortuously, a voting-rights advocate said that it looked like "Abraham Lincoln riding on a vacuum cleaner.") The result is a rigid, unchanging political terrain: once someone's in office, they almost can't be voted out. From 2002 to 2010, a slender 4 percent of incumbent New York state politicians lost. In 2010, nearly one out of five politicians didn't even have an opponent, because competitors realized there was no point. The game was thoroughly rigged.

"When things get this bad, democracy gets hijacked," Panagopoulos tells me. Indeed, New York has one of the most gridlocked and dysfunctional legislatures in the country, because rival parties have no incentive to cooperate. "It's a hidden issue that no one talks about how to fix."

The reason no one talks about it is simple: gerrymandering is a monstrously complex subject. To fix it, you need to analyze what's going on each city and suburb block, for hundreds of miles across an entire state, parsing an absolutely Olympian mountain of information (maps, databases, dense charts of voting data, and so on.) As a result, a professional class of map riggers has emerged, lushly compensated consultants hired by politicians to guarantee victory. The evils of this system are protected by the byzantine nature of the problem. Voters have little chance of figuring out why things are going so wrong, let alone of fixing it. "The average citizen throws up their arms and tunes out," Panagopoulos says.

Unless, as he realized, clever software could level the playing field.

In 2010, Panagopoulos teamed up with a group of academics and coders who had created District Builder, a mapping tool designed to let anyone redraw electoral districts. District Builder uses a point-and-click interface so clever and intuitive it's almost like a video game; the software actually makes gerrymandering *fun*. Panagopoulos and the team fed in all the maps and demographic information for New York State and put them online. Now any citizen could log in and redraw districts, making a particular constituency more or less competitive, stuffed with black voters or overflowing with Asian ones. The code would do the data crunching—performing, in effect, the behind-the-scenes work of those million-dollar experts. This would free up citizens to focus on the big ethical and moral questions: How can we make the district boundaries fair to everyone? What's the best way to force politicians to face real competition?

Panagopoulos organized a competition for college political science students. Could amateurs, with no expertise, produce fairer electoral districts?

Indeed they could. Over a two-month period, dozens of teams of students whipped the maps into shape using District Builder. The crazy, Abe-Lincoln-shaped districts were gone. Normal, square-shaped districts emerged. "Competitiveness" shot upward: when Panagopoulos examined the results, he saw that the districts had a much more equal blend of Republican and Democratic voters.

"It actually wasn't that difficult," marvels Alyssa Barnard, one of Panagopoulos's students. It's a sunny winter afternoon, and I've traveled up to Fordham University so she and her classmates could show off their skills. Barnard tells me she had focused specifically on the votes of ethnic minorities.

"I'd read a lot of stories about Hispanics being purposefully disadvantaged," she says, cracking open her laptop and pointing to the districts, each with simple shapes and clear boundaries. Teamed with a partner, she had needed only twenty hours to complete the process.

Another student, Tyrone Stevens, shows how easy it was to tweak things on-screen. "You can make things unfair pretty quickly," he jokes. Stevens circles Republican households, drags them into another district, and presto—the district abruptly turns bright blue, indicating he'd created a safe seat where Democrats would be unbeatable.

But what was most interesting was how the students' thinking changed. Playing with maps really was like experimenting with a video game: you poked around, doing experiments, gradually deducing how the system works. If I do *this*, what happens over *there*?

The more they messed around, the more the students' understanding of redistricting kicked in. While they'd read about gerrymandering and techniques to fix it, they'd had no intuitive grasp of the problem. But like carpenters who can sense when a piece of wood is too fragile by the vibrations of the nail, the students' use of the software gave them an almost tactile sense of electoral dynamics. When I looked at the map I saw a map. But when they looked, they now had *Matrix*-like vision—they saw the streams of convoluted demography that determined whether New York would turn into a rigid, dysfunctional mess or a competitive, healthy democracy. They could *see* the system, how it lived and breathed. "You're looking, and you realize, 'This doesn't have to be that way,'" as Stevens puts it.

For her part, Barnard says her hours of playing with the map led her to a surprising epiphany: It's actually easier to make a gerrymandered map than a fair one. Unbalanced districts aren't purely a result of greed and deceit. Even honest-minded attempts to alter a district can produce unexpectedly bad results, in a butterfly effect. You might try to make district A more fair by siphoning off households from district B, only to find it inadvertantly makes districts C and D and F *more* gerrymandered. It was a nuanced political analysis that stemmed from hands-on experience instead of abstraction.

I ask Barnard how long it would have taken her to redistrict the state if she hadn't had the software. What if she'd had to do it like in the old days—with pencil and paper, and stacks of population tables?

"Just as a citizen?" she says, raising an eyebrow. She shakes her head. "It wouldn't be possible."

Traditionally, "literacy" has primarily meant two things: being able to read and being able to write. That's still true, and it will remain true for a long time, because the written word remains an exquisitely flexible tool for formalizing and manipulating knowledge. And as I wrote in the last chapter, digital tools have created a renaissance in the written word.

But as the Fordham students realized, digital tools are giving us new and powerful ways of grappling with information. Technically, the students could have used pen and paper to work out the districting problem. But it would have been prohibitively slow, which is precisely why even very committed amateurs have rarely attempted it. A tool that let them play around with masses of data—a map interface laid on top of a huge database—lent them a far richer grasp of nuances, insights less easily available from reading about the subject. It gave them a new literacy.

Computational power isn't just changing the old literacies of reading and writing. It's creating *new* ones. This includes literacies in video, images, and data sets, forms of information that are becoming newly plastic. It wasn't long ago that only well-funded experts, mostly in large corporations or government agencies, could wield video or data as a mode of expression or analysis. Now those modes are used in everyday situations by ordinary people, as tools like District Builder become simple enough for anyone to use. This opens up our mental tool kit, helping us spy patterns in the world that were previously invisible to us.

Crunching data isn't new, of course. Statisticians have done it for years, as have government economists studying patterns in consumption data. But this wasn't considered something the average person ought to know how to do; public schools mostly teach rather rudimentary data analysis, like reading bar graphs and simple statistics. (Not that I'm denigrating bar graphs, which can lead to surprisingly powerful insights. My son's first-grade class charted the heights of everyone in the class, discovering as a result the principle of the bell curve.) For two reasons, it's not surprising that data analysis was, until now, a highly specialized activity. The first is that statistical software was usually too difficult for an untrained person to use. The second is subtler: There also wasn't a lot of data lying around that the average person particularly *wanted* to crunch.

All that is changing. First off, tools for crunching data are becoming much simpler, and even fun to use.

Consider one popular data-analysis tool, the word cloud. You've probably seen or even used one of these Web tools, such as Wordle. To make a Wordle, you cut and paste a piece of text into the generator, which analyzes the frequency with which each word in the text occurs. The program then creates a "cloud" image, with the most frequent words looming in huge font sizes and the less frequent ones in increasingly smaller sizes. This turns a complicated metric—*what words and concepts define this piece of writing?*—into something anyone can evaluate at a glance. Wordles became famous in the 2008 U.S. presidential election, when pundits began generating word clouds from the speeches of the rival candidates, John McCain and Barack Obama. One blogger, Heather Williams, produced a sharp visual contrast: when she analyzed stump speeches in June 2008, McCain's cloud was dominated by "oil" and "energy." It was instantly clear that those were his twin focuses, above all else: America needed to drill for more oil, right now! In contrast, Obama's main words were "children," "Americans," "make," "care," and "need." But the really striking thing is that Obama's cloud didn't have very many big-font, standout words at all. His speeches didn't have a single, tight, overwhelming focus in the way that McCain's did. In retrospect, these little Wordles may suggest a sophisticated point about that American election: voters were more interested in the candidate who talked about a broader array of issues.

Teachers have begun using Wordles to help students improve their writing by teasing out bad habits in their prose. For example, Melissa Baralt, a professor of Spanish applied linguistics at Florida International University, ran an experiment in which students input the first draft of their Spanish-language essays into a Wordle. The first thing they noticed was that their vocabulary was too repetitive. A few words and phrases—like *mucho* ("many") and *pienso que* ("I think that")—loomed large. They rewrote their essays three times, trying to use a more diverse vocabulary, and the improvements were immediately obvious. By the final draft, the vocabulary usage of the whole class had broadened by more than 30 percent.

Other teachers have used Wordles to help students analyze literature. Erin Templeton, an English professor in South Carolina, found that students often had trouble reading analytically. They were good at plot summary; they could describe what *happened* accurately. But they couldn't analyze books in more abstract ways—delving into larger themes, the author's use of language, or the book's literary style. So Templeton began creating word clouds from F. Scott Fitzgerald's books, including *This Side of Paradise*. Odd words would pop out (such as "glue" or "nervous"), and the students began thinking about why certain nonobvious phrases dominated a passage.

"I thought maybe it would help, but I was stunned at how much it worked," Templeton tells me. "They finally understood that I'm not asking you what this *says*, I'm asking you how it *works*.'"

A Wordle is a very simple—even crude—example of a data tool. But the sheer number of such tools is erupting because of a second trend: There's simply more data lying around. A decade's worth of e-mail, text messages, photos, and health information can provide fascinating glimpses into hidden contours of your life, if you're able to seek patterns in the noise.

This is the experience of the users of popular self-tracking tools, like GPS-enabled watches or step-tracking pedometers like the Fitbit, which offers an at-a-glance view of how much your performance is improving. One of the side effects of regularly seeing this sort of chart, it turns out, is that astute users begin to notice subtle trends in their health. Some users of the Nike+ step-tracking device have found that by looking at the trend lines in their running performance, they can spot incipient muscle strain—even before they physically feel it. Other self-trackers have used wristwatch heart-rate monitors to predict the onset of a cold, allowing them to take extra care of their health to avoid it.

Or there's the story of Alexandra Carmichael, who is the community director for the Quantified Self Web site, and organizes meetups of people worldwide who like to track their personal data. Carmichael had long wrestled with depression, so she wondered if she, too, could learn more about it by visualizing its ebb and flow. She heard about Moodscope, an online tool that enables users to report on their daily moods. After six months of using it, she analyzed her chart and discovered, to her surprise, that her state of mind fluctuated far more than she realized. The chart looked like a jagged mountain peak. Indeed, her highest point in the six months was the day she walked a marathon—and the lowest point came a mere three days later, when she noted that she was "jittery, lonely, sad, overwhelmed by work/family demands." Before, Carmichael had believed her depression moved in slower cycles, but it was actually quite volatile. This became a useful piece of self-knowledge. "Once I saw the pattern of my mood going up and down so much, I started having a sense on down days that all I had to do was wait and my mood would go back up," as she wrote about the experience. "So the down days didn't seem as dark as they used to."

The welter of things we can gather data on is going to expand dramatically in the years to come. Just as Moore's Law drove down the price of computer chips, sensors are getting cheaper and cheaper. Apple and other companies have already patented technologies that let your mobile phone sense your heart rate when you hold it, which would allow the phone to read your mood; it could know when you're excited, annoyed, or maybe just too busy to take a call or text. Motion-sensing devices like the Kinect for Xbox can similarly read your body language, and even webcams and the video cameras in mobile phones provide evidence of your affect. Dale Lane, an IBM programmer who lives in the United Kingdom, used a webcam mounted on his TV to track his expressions while he was in the living room. This produced some delightful epiphanies: He discovered that he smiled most often during the opening part of the BBC show *Mock the Week*, whereas he mostly looked angry while playing his Xbox games online. He also found that he was about five times more likely to smile when his two young children were in the room with him than when he was alone. (And when he tracked himself while sitting and working at his MacBook, he discovered he looked pretty grim: "Do I not smile very much?")

Will everybody actually *want* to develop data literacy? Probably not, just as not everyone wants to develop their skill at writing or pencil drawing. Still, the number of people who already find data visualization enjoyable is surprising. Five years ago, Fernanda Viégas and Martin Wattenberg, two IBM scientists (they now work for Google), launched Many Eyes, a Web site that offered a set of easy tools for creating colorful charts, bubble graphs, and word clouds. "We didn't know what to expect," Viégas tells me, but pretty soon ten thousand users had stormed in and began excitedly creating visualizations of almost everything. When I visit their office in Boston, the two pull up examples to show me. One woman had charted her company's sales, trying to find patterns governing its success; others crunched statistics on sports teams to discover unexpectedly powerful players.

Many Eyes became a locus of public thinking, too: the users began having conversations about each other's charts, critiquing and improving them. One user who was interested in levels of violence in the twentieth century created a chart plotting how casualties steadily climbed during the wars and ethnic purges of the last one hundred years. It made for a pretty bleak picture, which was precisely the user's point: Life's gotten more violent, right? But soon other users pointed out that the chart was misleading. It didn't account for the boom in population. So the original chart maker added in a trend line showing population, updated the chart, and the picture changed: Now you could see that violence had increased more slowly than population.

"It elevated the level of discussion," Wattenberg says. "It's sort of like when writing was invented, and people were saying, 'Is this going to be something special that only the priestly class can do?'"

"It becomes this gateway for young people all the way to adulthood to get in touch with statistics," Viégas adds.

"We'll know it's successful," Wattenberg jokes, "when visualization becomes boring." It's true: Tools for thought often are most powerful when they become second nature, thus somewhat invisible. Writing transforms thinking only after the point where you no longer have to struggle with the act of writing itself.

If you want proof that data visualization is entering the mainstream, it's there in online pop culture. Some of the biggest viral hits in recent years have been witty data crunches from odd, unexpected sources. Arthur Buxton, a young British Web designer, analyzed the color usage in Van Gogh's major paintings and transformed them into pie charts, challenging viewers to figure out which was which. A group of Christian data nerds did a "sentiment analysis" of the Bible, using algorithms that determine whether a piece of text contains positive or negative language. ("Things start off well with creation, turn negative with Job and the patriarchs, improve again with Moses. . . . In the New Testament, things start off fine with Jesus, then quickly turn negative as opposition to his message grows.") A professor of English used network-mapping software to analyze how characters interact in *Hamlet* and produced a map that uncovered some revealing patterns. (If you look at all the characters that speak to both Hamlet and Claudius, only two survive the play; and Rosencrantz and Guildenstern never speak to each other.) And everywhere online there are vicious, delightful parodies of Power-Point presentations, the cesspool of data visualization that Microsoft has visited upon the earth.

PowerPoint, indeed, is a cautionary tale in our emerging data literacy. It shows that tools matter: Good ones help us think well and bad ones do the opposite. Ever since it was first released in 1990, PowerPoint has become an omnipresent tool for showing charts and info during corporate presentations. But the design expert Edward Tufte argues that PowerPoint also tends to *mangle* data. In 2003, Tufte analyzed 217 data graphics made using PowerPoint (taken specifically from textbooks on how to use it). He discovered that because PowerPoint slides are low resolution and designed to be projected on a wall and read at a distance, the tool encourages people to be overly reductive. As a point of comparison, he looked at the gold standard of everyday charts—the data-rich ones you'd see in newspapers like *The New York Times*. Tufte found that the average *New York Times* chart contained 120 elements; with that many pieces of data, a reader can spy interesting patterns. PowerPoint slides were so primitive that the average chart contained only twelve elements. As a bleak point of comparison, the average chart in *Pravda*—back when it was an official instrument of Soviet propaganda in the early 1980s—contained five elements. PowerPoint is so limiting a tool, Tufte concluded, that it nudges users toward an almost Soviet level of obfuscation. "Doing a bit better than *Pravda*," he wrote, "is not good enough."

If we want to have a data-literate future, we need to figure out which tools serve our thinking and which ones don't. As with all cognitive shifts, that requires work.

In the fall of 2011, the AMC series *Breaking Bad* was in its fourth acclaimed season, with a devoted, nearly fanatical, audience. Bryan Cranston plays Walt, a meek chemistry teacher who launches a new career cooking up crystal meth. Originally Walt does this out of desperation: He's got cancer and needs to pay for his treatment. But even after his cancer recedes and he's rolling in drug money, he discovers that he's addicted to the danger of being a drug baron. Viewers loved *Breaking Bad*'s moral ambiguity, its wild performances, and the cinematic camerawork that tracked Walt's creeping descent into violence.

But toward the end of the season, a mystery emerged. The son of a central character had taken mysteriously ill with violent, flulike symptoms, and nobody could figure out what had happened. What was going on? Fans swarmed onto discussion boards to argue about it.

A few days after the baffling episode aired, a *Breaking Bad* fan named jcham979 posted a six-minute video to YouTube that offered an explanation. Jcham979's theory? Walt had done it. He'd poisoned the the boy in an attempt to frame Walt's archenemy. Walt was now so evil that he was willing to risk the life of an innocent boy just to maintain his drug empire.

But the really remarkable thing was *how* jcham979 articulated his theory: He created a masterfully crafted video. He combed through the episodes of *Breaking Bad*, locating moments that seemed innocuous at first viewing but that, if you looked really closely, contained hidden clues. Then he edited them together into a sequence that revealed the plot, adding bits of on-screen text to point out details most viewers had missed. For example, jcham979 replayed a scene in which Walt, recently beaten up by his archrival's goons, stares morosely at a flower in his garden. The flower, jcham979 noted, was a lily of the valley—"a toxic plant that is not much of a danger to adults . . . but has a much stronger effect on children," including producing lethal symptoms that resemble the flu. In another section of his video, jcham979 froze frames from a fast-paced scene to analyze, at practically the millisecond level, how Walt planted fake evidence on a friend. Jcham979 highlighted moments that were barely visible to ordinary viewers.

It was a dazzling act of analysis. It also turned out to be right: when the final episode of *Breaking Bad* aired days later, jcham979's theory held up. Walt *had* done it, using lily of the valley to poison the boy. By scrutinizing the scenes with such frame-by-frame attentiveness, jcham979 had picked up complexities other fans hadn't, presenting his evidence in a brilliant, visual way.

Until recently, of course, it wasn't possible to do this with the moving image. In the decades leading up to the twenty-first century, it was difficult to impossible to parse in such detail anything that appeared in movies or on TV. You couldn't easily rewatch it. And you certainly couldn't cut and paste together your own video as a method of argument.

The slippery nature of the moving image had long troubled cultural critics. In his 1985 book *Amusing Ourselves to Death*, the cultural critic Neil Postman bemoaned the passing of the age of print. Print, he wrote, allowed for parsing and rereading. Electronic media just slid past you into an eternal memory hole. "The eye never rests, always has something new to see," Postman complained. While burning a book was heresy, electronic media, ever since Samuel Morse's telegraph, was structurally impermanent: "*the telegraph demands that we burn its* contents." This meant that the moving image could really never be used for careful thinking, the way we'd long used writing and reading.

For the time he was writing, Postman was absolutely right. But by the mid-2000s, things began to change dramatically. A riot of new tools emerged: hard-disk recorders like TiVo let viewers pause TV shows and scan them on a frame-by-frame basis, even as they were being aired. Digital video-editing tools became simple enough for children to use. Digital video cameras became so cheap they were incorporated into nearly every phone and laptop. And video-hosting sites like YouTube made broadcasting the moving image as trivial as blogs had made broadcasting print. In essence, we've seen a whipsaw transformation as tools emerged that allowed for a new level of literacy in the moving image—with TV itself as the first subject to be dissected.

When the TV show *Lost* went on the air in 2004, fans (me included) were baffled by the plot. Smoke monsters? Polar bears on a tropical island? A sequence of six numbers (4, 8, 15, 16, 23, 42) that kept cropping up? It was a crazy puzzle, begging its viewers to make sense of it. So that's precisely what the fans did. They began devoutly recording episodes, capturing screenshots that seemed to contain clues, and arguing heatedly over what they meant. (Or what they didn't mean, when it became clear that the directors themselves had failed to close innumerable plot holes.) In 2012, the second season of the BBC show *Sherlock*—a modern refiguring of Sherlock Holmes—ended with Holmes hurling himself off the roof of a building and his bloody body being picked up off the ground below. It was a ruse; the episode ends with a shot showing Holmes is still alive. Fans immediately flooded online, posting screenshots and clips supporting theories explaining how the elaborate trick was pulled off. I spent one evening reading a ten-thousand-word-long discussion thread that fans had filled with screen grabs and clip-fueled evidence. One had even constructed a Google Map of the neighborhood where the BBC had shot, so that you could shift to "street view" and wander virtually around the building where Holmes jumped, the better to deduce how he might have survived.

In some respects, this frame-by-frame study of TV takes us back to the origins of the moving image. In 1878, the photographer Eadweard Muybridge helped settle an issue over which horse racing aficionados had long bickered: When a horse is galloping, is there a moment when all four hooves are simultaneously off the ground? Horses gallop so quickly that you can't clearly perceive this with the unaided eye. So Muybridge invented rapid-fire photography, taking a series of pictures of a galloping horse mere milliseconds from each other, producing a flip-book-like sequence. Sure enough, he proved that the four legs of a galloping horse do leave the ground all at once. Better yet, he invented the first animated video player, the zoopraxiscope, to play the images in succession. When he took it on the road, crowds were fascinated by Muybridge's ability to slow down, speed up, and freeze-frame the horse. The moving image emerged, in a sense, from our desire to scrutinize the evanescent visual moments of daily life. Our modern digital tools are returning it to those roots.

The new scrutiny of TV isn't limited to pop-culture shows. It's also transforming politics. No sooner does a politician make an appearance on television than political junkies and activists have scanned in the video, uploaded it, clipped out statements, and begun debating what was said. This new form of analysis has reached its apotheosis in programs like *The Daily Show*. Despite being putatively a comedy show on a comedy channel, *The Daily Show* has become one of the most-respected news sources among young people in part because it's so deeply literate in inspecting TV utterances. A common technique is catching politicians in acts of staggering hypocrisy—running a clip of something they said today, followed instantly by a clip from weeks or months earlier in which they said precisely the opposite. The show's fluency developed in the 2000s as a result of technology advances: The producers built a "nine-foot-tall rack" of hard-disk recorders and monitors that pick up broadcasts on oodles of stations all day long, for later scrutiny.

In a few years, it's likely this style of omnivorous recording will be available to anyone who wants it. In 2012, a British inventor unveiled the Promise TV box, which records the broadcasts of more than sixty TV channels simultaneously, twenty-four hours a day, for a week, and allows for hyperlinking to individual segments inside any show. Costing more than three thousand dollars, it was a ridiculously expensive gadget, but the ever-plummeting cost of data storage will probably make that scale of recording cheap, even ominpresent, in less than a decade. With this type of tool, Postman's worries become inverted. The moving image not only doesn't vanish; it never goes away. The challenge is no longer catching it but making use of it—our age-old problem of navigating increasingly huger libraries and ever-stranger new floods of information. The challenge is daunting, but it's one we're historically excellent at tackling.

The new literacy of video works in reverse, too. The moving image isn't just easier to parse and critique. As jcham979 showed, editing software and online distribution have turned video into a flexible format for communication, a way to speak one's ideas to the world.

The word processor did the same thing, thirty years ago, to text. Most older people have forgotten this, and most younger people never experienced it, but before the word processor went mainstream in the 1980s, professionally editing text—moving ideas around, sorting your thoughts—was a surprisingly cumbersome chore. Cutting and pasting was not metaphoric but literal. Editors at newspapers would take a typed-up story and edit it by cutting it into sentences, rearranging them, and laboriously pasting them into a new order on a fresh page. That glacial pace had its advantages. There was a meditative quality to shuffling the scraps around, as various gray-haired newspaper editors, hovering on the edges of their retirements, informed me when I became a copy-editing newspaper intern in 1994. Shuffling paragraphs could spark new connections and ideas; plus, working in such a forcibly slow fashion can push your brain into connective, dreamy modes. This is precisely why it's still a good idea, even today, to step away from your laptop and work out a problem on paper: cognitive diversity, again.

But adding the speed and flexibility of the word processor to our tool kit has been a huge boon to thought. In his visionary 1962 essay "Augmenting Human Intellect," Douglas Engelbart—a pioneer of the computer mouse and the files-and-folders graphical desktop—dreamed of the way cutting and pasting would one day transform our thinking. Writers equipped with electronic word processors, he prophesied, would become superior brainstormers, able to jot down ideas as rapidly as they came, then slowly hone them at leisure. We wouldn't lose our trains of thought so easily. Stray ideas could be preserved. "You can integrate your new ideas more easily, and thus harness your creativity more continuously, if you can quickly and flexibly change your working record," he wrote. Engelbart was right. Studies show that when we use word processors we're more iterative; we write our first drafts more quickly but then spend more time revising them. We do more of our thinking *on the page*, as Richard Feynman would have it, externalizing our thoughts first so that we can ponder them in front of us. Word processors and desktop publishing created a grassroots renaissance in print by making it easier for people to publish tiny, obsessive niche "zines," the first blast of the DIY creativity that would later become routine on the Web.

This same process is now beginning to apply to video. That's what happens when you reduce a personal video camera, in the space of two decades, from a nine-hundred-dollar gadget the size of a shoebox to a Chiclet-sized component costing less than ten dollars and shoved into every laptop and phone. Video sites such as YouTube are now crammed full of videos made by ordinary people to communicate. Time lapse has become a popular meditative technique, as with the young blogger clickflashwhirr, who took a picture of her face every day for five and a half years, then blurred it into a mesmerizing minute-long spectacle of memento mori. There are "supercut" videos that combine dozens of common tropes in movies or TV (every time the cook Rachael Ray says "mmmm" while tasting food; glamorous hackers wearing leather) as a way of mocking, and celebrating, clichés. There are endless mash-up artists, like the YouTube creator Kutiman, who makes new compositions by taking tiny snippets of different songs and blending them into a single piece—or Jonathan McIntosh, who neatly critiqued the creepy, stalkerlike behavior of the lead vampire in *Twilight* by remixing his scenes with those from the avowedly feminist *Buffy the Vampire Slayer*. There is a constant flood of live citizen news; during the Arab Spring, when journalists were banned from many of the countries in which protests were taking place, government crackdowns were documented primarily by the protesters themselves. And there are conversational forms emerging, like the response video, a type of commentary that has essentially no analogue from the pre-Internet video universe: People commenting on a video by recording their own response, which itself gets responded to, and on and on.

What's striking about these videos is how *weird* many of them are aesthetically. The riotous amateur quality of much online video is reminiscent of the hallucinogenic short films that were made in the late nineteenth century, when film was brand-new and no one knew how to use it. Eventually Hollywood and TV developed a visual language—the close-up, the tracking shot, the panning shot. They developed genres, like the western movie and the TV-show interview. In many cases, amateurs using video ape precisely those shots and genres; they try to make their videos look "professional," which is to say, like existing commercial TV and film. But they also do things that violate the dictates of traditional video, because the ease of shooting and editing allows for odd new experiments with the form. Digital video has made the moving image much stranger.

Consider the confessional shot, used in most response videos—someone staring right into the video, talking casually and off the cuff, usually seated in their bedroom or living room, often with lousy lighting and worse audio. It is a shot that was aesthetically unimaginable before the digital age. You'd rarely find it a traditional movie, because it's too ugly. You wouldn't even find it often on TV news, where authority comes from being well lit, made up, and seated behind a desk. Even documentaries rarely employed such grungy-looking, talking-to-the-shaky-cam visuals. (The closest mass parallel was the "confessional" shots in the ur–reality show *The Real World*.) Yet measured in sheer volume, I'd guess that the talking-into-the-webcam video is now the most common shot in the entire 150-year history of the moving image. When you put a webcam on every computer, that's what people are going to do: Turn it on and talk.

They'll also collaborate with one another. Consider an experiment conducted by MadV, a YouTube user whose videos often showed him performing magic tricks while wearing a Guy Fawkes mask. MadV posted a forty-one-second video of himself holding up his hand, on which he'd scrawled the phrase "One World," then flashed a message on-screen urging viewers to respond. Within hours, people worldwide posted videos of themselves with replies written on their palms—like "Don't quit!," "Tread gently," "Carpe Diem," and my favorite: "They could be gone tomorrow!" Each video was barely a few seconds long, but there were a lot of them: MadV received two thousand such video responses, the most in YouTube's history. He stitched dozens of the most provocative ones into a montage—a stream of people communicating with words on their hands, a video that possessed a strange emotional power. Yet it's hard to say exactly what his video *was*. A conversation? A documentary? Some new type of poetry?

Such eclectic experimentation is crucial in expanding the language of the moving image, unlocking its potential as a mode of expression. To see where things are headed, it would be best to pay attention to these seemingly trivial or even crude amateur experiments, because it's often in the realm of the amateur where the truly new is possible; amateurs are exempt from the commercial pressures that have locked professional TV, film, and industrial video into its narrow palette of genres and shots. Mind you, it's also true that today's amateurs probably aren't experimenting aggressively enough. Most are still trying to mimic the language of traditional video, producing material that apes the form and content of TV and movies. We're in the transitional moment that, as Marshall McLuhan points out, occurs at the beginning of every new medium. At first, we have no idea what to do with it—so we tend to repeat the style of the *last* major media. When printing first made it possible to mass-produce novels, authors tended to recycle the older tropes of letter writing, mock-heroic poetry, and nonfiction. It took time for the modern psychological novel to develop. Early TV imitated radio and vaudeville as well as theater and movie serials; ambitious serialized storytelling, exemplified by shows like *Breaking Bad*, took decades to emerge. Online video will similarly take a long time, probably decades, to evolve away from the established conventions of commercial TV and film. And of course, most of these early amateur experiments will be off base and idiotic and will fail miserably. That's why the truly innovative stuff on YouTube or other video sites is still in the tiny minority.

Another key area of video literacy is our growing ability to capture ideas that are difficult to represent in text. Consider, for example, the explosion of how-to videos. These days, if you need to learn how to accomplish a task that requires real-world physical skills—knitting, engine repair, geometry, parkour, cooking—amateur how-to videos have become the best way to grasp it. That's because complex physical skills are often incredibly hard to describe in words. Two years ago, I wanted to learn how to build guitars, which required the use of dangerous high-speed tools. Since I didn't want to sever my fingers, I tried reading some training manuals on the subject to almost no effect. You can't learn physical skills that way. Instead, I turned to the sprawling universe of amateur how-to videos; they were far more visually eloquent, and in about five minutes I absorbed more than I'd gotten from an hour or two of reading. It is no accident that the "maker" movement, a worldwide collection of nerds trying to learn and teach everyday mechanical and electronic know-how, has arisen in the age of easy video documentation. If you want to know how to build something, seeing it happen is crucial. These complex physical actions and bits of problem solving *can* be discussed in text, but as Nils Nilsson, a pioneer in artificial intelligence at Stanford, once told me, it's not necessarily easy, or even the way our minds work. "You can describe in words how to swing a golf club," he added. "But can that really tell you how to do it?"

Video literacy is also evolving rapidly, because a global audience can quickly see and replicate successful experiments. Chris Anderson, the curator of the TED conference, notes that when TED began posting free video of its top-ranked talks, the polish and dramatic flourish of presentations shot upward (to an almost freakish degree, I'd add). Why? Because the speakers were able to study the previous years' videos, deducing techniques to better craft their own talk. Similarly, Hunter Walk, former manager of product development at Google (the owner of YouTube), has watched as innovative editing tricks pass from one user to another. "People play with different types of filmmaking styles. They populate, spread, carry forward faster," he tells me. Hollywood evolved its visual language at a comparatively slow pace, he notes, because the production cycle of a movie is measured in years. But once a single person figured out how to produce "tilt shift" footage on YouTube—which makes regular-sized buildings and people look toy-sized—others began copying the effect within days. "It's a community sharing of information," Walk says. "You have the ability to tell people how you did it, or for them to reverse engineer it. So the information gets out in a global fashion, and you no longer have this creeping, twelve-month spread. And everybody is in this duality of being a creator and a consumer. So you're essentially riffing on everybody else's work. It's not just repurposing people's content. It's repurposing and expanding people's *technique*." There was recently a boomlet in animated gifs—a small handful of frames scraped from a TV show or a video, showing a half second of screen time, like *Star Trek*'s Captain Picard grimacing while "face-palming," repeating endlessly in a zoopraxiscopic loop: Muybridge's invention reborn as a new type of emoticon. The fad may soon vanish, but it'll be replaced by some odd new experiment in using the moving image.

Still, for video to really advance as a medium for thinking, there's one major shift that will have to occur: We'll need to begin using it to communicate with ourselves.

Right now, of course, the moving image is inextricably tied to the idea of broadcasting to an audience. Why would you record video except to show to someone else? But think of the various cognitive roles of other flexible media, such as the written word. When we write, we're not always communicating with others. Frequently we're communicating with ourselves, using pen and paper or a word processor to organize our ideas, to parse and reparse information. This personal, inward-focused use of writing took a long time to evolve. That's because it was once expensive and difficult to write; plentiful blank paper and reliable pens were rare treasures for nonelites until the late nineteenth and even early twentieth century. (While fighting the British, George Washington's generals complained they didn't have enough paper even for crucial letters. "Excuse these scraps of paper; necessity obliges me to use them, having no other fit to write on," one wrote.) But as paper and writing implements became cheaper and more abundant, we began using text to record our ideas not just for others but for ourselves. In today's industrialized countries, paper is so omnipresent that we can deploy it in a throwaway fashion—using Post-it notes for short-term memory, to help us manage complex tasks.

So the question becomes: What will be the *video* equivalent of a Post-it note? What innovations will make it as flexible, dash-offable, and refindable as words on paper? We can catch a hazy glimpse of the future in current lab experiments. Engineers are developing algorithms that let you find specific moments in video, much as you search for and find words inside a long text document. Others are developing desk-sized screens that let you compare snippets of video the way you compare print documents, or that recognize the content of similar scenes—making the moving image archivable and refindable in the way the library made the book, and the Web the article.

In the opening sequence of *Minority Report*, Tom Cruise stands in front of a huge holographic display, arranging different chunks of video. He rewinds and studies the snippets, deducing the identity and location of a murderer. Reviewers cooed over the enormous, wall-sized screen. But what struck me was the cognitive task at hand. Cruise's character was using video to solve a problem—to talk to himself and help himself think.

If you want to see what these literacies will look like as they mature, examine the one that's come the furthest: photography. Consider the case of Alexei Navalny.

Navalny is a Russian blogger and democracy advocate, as well as a critic of Russian president Vladimir Putin. He coined the phrase "a party of swindlers and thieves," and the idiom went viral, chanted by pro-democracy protesters in 2011. Eventually the authorities had enough and threw Navalny in prison. But by then, Navalny had become a national hero, and demonstrators poured into the streets until the government was forced to release the blogger.

As payback, the government and its supporters decided to try to discredit Navalny. A pro-government newspaper crafted a fake photo that showed Navalny laughing next to an exiled billionaire, a supposed enemy of Russia. Then they published it with an explosive cutline: Navalny, they claimed, wasn't a hero. He was just another crook on the take!

But the scheme quickly unraveled. Navalny's supporters passed the photo to him, and he posted it on his site to decry the fakery. Within hours, the photographer who had taken the original undoctored photo posted that picture online, too, proving that Navalny had been standing next to a completely innocuous person, and the exiled billionaire had been photoshopped in. Then the fun began. Navalny's supporters turned the doctored photo into a playful meme. Protesters began creating their own manipulated versions of the photo to mock the government's ludicrous attempt at character assassination. They photoshopped in various silly figures, posing Navalny next to a space alien, Arnold Schwarzenegger, and Hitler. The meme kept galloping along online, mocking the government anew with each variation. The joke wasn't just that the government and its supporters were corrupt. It was that they were *inept* for presuming forged photos would go undetected in a visually literate age.

It was, of course, Soviet leader Joseph Stalin who pioneered the use of photomanipulation as a political tool. He loved to create faked photos of his rivals in incriminating positions and to erase from photos ministers he'd had killed. Stalin's photomanipulations helped inspire the sinister Ministry of Truth in George Orwell's *1984*. Orwell feared that when powerful people alter the documentation of history, they change reality, since the population will be unable to detect the truth.

But Orwell wrote before the advent of Photoshop as a folk art. Stalinesque alterations used to require experts to execute; in the intervening decades, though, they've become easy enough for elementary-school kids. Even if you can't afford Photoshop itself, free tools abound, including online sites offering point-and-click apps for cleaning up, tweaking, and altering one's own pictures. In the mid-twentieth century, Orwell and Stalin correctly assumed a level of public ignorance about the plasticity of photos. Putin's supporters, to their regret, assumed that level of ignorance still exists. But the average person today is not only much more literate in photography, she's also networked, which makes her smarter yet: Whereas one person might have trouble figuring out whether a photo is fake, several thousand people working together can untangle the deception in hours, even minutes.

Repressive governments worldwide, rather hilariously, seem to be having trouble grasping this shift. In 2008, as a show of military might, Iran published a photo showing a simultaneous launch of four missiles. The photo ran on the front pages of the *Los Angeles Times*, the *Chicago Tribune*, and the Web site of *The New York Times*, sending pundits into a frenzy: Iran's missile capabilities were much more advanced than we thought! But many online bloggers, schooled in Photoshop, smelled a rat. Within hours, they'd pointed out the photo's fakery. Iran had launched only three rockets and had pasted in the fourth to make their feat look more fearsome. Newspapers and their expert photo editors didn't notice it. Online crowds of visually literate readers did.

Of all the new literacies, photography is the most advanced simply because the toolset has been in the hands of amateurs the longest. Kodak introduced its Brownie camera in 1900, promoting it specifically as a tool for the masses. The century-long experiment with personal photography that followed gave many people a rough sense of how a shot could be framed to create an aesthetic effect. The speed of experimentation accelerated dramatically in the 2000s, as digital cameras got baked into almost every smartphone; no longer did you have to wait for days to learn if your attempt at a new style of shot was successful. Photomanipulation, like googling and Xeroxing, has even given birth to its own piece of slang: an altered photo has been "shopped" (or the funnier "shooped").

Ten years ago, photo editors at newspapers worried aloud that "truth" in photos was dying, but the reverse is happening. Corporate photomanipulations are increasingly scrutinized and mocked for their inauthenticity. This has been particularly true in the case of women's magazines. Eagle-eyed readers now parse pictures of models, pointing out necks that have been enlongated and limbs made slender, as editors attempt to craft a feminine body that not only rarely exists, but *can't* exist. Much like dictators, corporations are pretty unsettled by the scrutiny. When Xeni Jardin of the *Boing Boing* blog poked fun at a Ralph Lauren advertisement in which a model's waist had been shrunk to nearly Popsicle-stick dimensions—"Dude, her head's bigger than her pelvis," Jardin wrote—company executives became so enraged they tried to have *Boing Boing*'s post removed with a copyright takedown notice. This led to yet more mockery, as commenters hunted down original images of the model, Filippa Hamilton, and compared her unaltered body (skinny, but within a human range) to the extraterrestrial manipulations of Ralph Lauren.

The precise moment that text-picture memes went mainstream was probably with the LOLcat—a picture of a cute animal layered with intentionally illiterate text. As the joke spread across the globe, pundits soon began castigating the LOLcat as another example of the dumbing down of digital culture.

But LOLcat-crafting skills can become quite powerful when

applied to other areas—even as political speech. In China, visual creations have been crucial in subverting the government's censorship regime. Why? Because pictures are less easily autocensored than text. The Chinese government requires local Internet firms to automatically scan for banned phrases and delete them. If you're a user of the major Weibo services—popular microblogging platforms similar to Twitter—and you try write about Tiananmen Square or Falun Gong, your post is likely to be caught and deleted. But today's software is still terrible at scanning images; even if it can autodetect a particular image, it can't figure out if you're using it to make an ironic point. That's an act of deep-meaning perception that machines can't match, for now anyway.

So to make their political points, Chinese Weibo users instead began posting witty, caustic graphics, mash-ups, and altered images. In the summer of 2011, two trains collided near Wenzhou, a city near Shanghai, and a reported forty Chinese died. Critics argued that the crash was a result of the Chinese government's lax safety standards, and they flooded Weibo services with postings attacking the government. The posts that stayed up uncensored the longest, outwitting the speech filters, were visual: the Chinese rail service logo altered to be dripping blood; a sea of Chinese citizens holding up a damaged rail car; a picture of a train ticket with the destination photoshopped to be the Chinese character for "hell." "This is what street art would look like on social media," as the American artist An Xiao Mina, living in China at the time of the crash, wrote.

Photographic tools have arguably reached the Post-it note stage, insofar as people regularly snap pictures as memory aids and retrieve them on the fly using tools like Evernote. We're also increasingly using photography to re-see the world, with the rise of "filtering" culture. You've likely seen this in services such as Instagram, which became popular in 2012: Users take pictures, then add a filter that tweaks the color composition, often highlighting details that aren't normally visible. Cultural pundits have tended to dismiss it as mere hipster nostalgia—the fun of making photos look old. And yet there's a deeper aspect to filtering, in that it allows amateurs to experiment with the perceptual texture of pictures in a way that only professional photographers could in the days of analog film. Nature photographers of the twentieth century loved Fujifilm's Velvia film because of how it saturated colors and offered stark contrast: Reproduced in the pages of *National Geographic*, Velvia photography changed the way a generation of North Americans appreciated nature. "Filtering, and otherwise bending visual light to our creative tendencies, has been part of professional photography since the beginning," Instagram cofounder Kevin Systrom tells me in an e-mail. "All Instagram did was take the creative tools that the pros have been using and put them in the hands of the masses." It develops the skill that Edward Weston, the pioneering photographer of the early twentieth century, called "seeing photographically"—using the camera for "augmenting the eye, seeing more than the eye sees, exaggerating details . . . a vital new way of seeing." With practice and training, many more people can develop photographic sight.

In fact, many artists of the 1930s predicted that photography would become a commonplace tool for thought. As László Moholy-Nagy, the Bauhaus designer, wrote: "The illiterate of the future will be ignorant of pen and camera alike."

To be clear, I'm not predicting that the written word, our oldest mass literacy, will disappear. In fact, it's likely to remain the go-to mode for expression. But as we develop ever more new modes for expressing our ideas and recording knowlededge, the challenge will be figuring out *when* to use *which* form. When is text the best way to make a point? When is the moving image? Or photos, manipulations, data visualizations? Each is useful for some types of thinking and awkward for others.

For years, commercial journalism struggled with the problems of using one mode to talk about everything. If you started a TV station, you had to tell every story with the moving image, even when it clearly wasn't suited for slow-moving, nonvisual issues (like multigenerational cycles of poverty). If you started a radio station, you had to tell every story with sound, even when that was less useful (as when discussing a visual artist). And if you had a newspaper, you used text and photos for everything, even when they didn't work very well (as with debating an issue, where the live, to-and-fro interplay of guests arguing on radio can work better).

But these silos break down in digital environments, where any news organization can deploy any tool. Right now, media sources have responded in a confused fashion, by reporting a story in *every* mode: a newspaper publishes a print story and then also does a (somewhat inept) podcast; a TV station runs a segment and also writes a (somewhat inept) news story. In the future, such redundancy will likely fade as journalists realize the trick is to figure out which medium best suits a topic.

And newer literacies will continue to come down the pike. Each year potential tools for thought are introduced, including some rather unexpected ones. One already emerging on the margins is 3-D design—using 3-D printers, machines that quickly produce physical copies of what you conjure on-screen.

This technology has been around for twenty-five years but was used exclusively by high-end designers because a printer cost tens of thousands of dollars. In the last five years, though, it has abruptly become a DIY tool: Start-ups like MakerBot sell home 3-D printers starting at eighteen hundred dollars, and toy companies are aiming to make even cheaper ones. When I visited a local "Maker Faire," an event where amateur inventors show off their creations, I found a row of 3-D printers quietly printing out various objects. One was fabricating gears for a funky-looking clock. The inventor who'd designed the clock handed me a freshly printed gear, blue, spiky, and pretty. "It's basically the same material they make Lego bricks out of, so it's pretty tough," he noted. Kids swarmed around to design their own objects. It's these kids, who have free time and no preconceptions, who will likely be the ones to domesticate 3-D printing, just as they were the first to domesticate computers, printers, Photoshop, and video-editing software. Over at sites like Shapeways or Ponoko or Thingiverse, creators post designs for everything from iPad racks to Rubik's Cube–like puzzles. Many are "open," which means anyone can download them, customize them, and print a copy themselves—learning gradually by remixing existing works, much as we learn to write by copying or imitating others.

What literacy will 3-D printing offer? How will it help us think in new ways? By making the physical world plastic, it could usher in a new phase in design thinking. 3-D printers allow us to meditate on physical solutions to physical questions. Julian Bleecker, a designer for Nokia, once told me about a colleague who was particularly fluent at thinking with Nokia's 3-D printers. "I'd be describing an idea for a new tool, a new invention, and couple of hours later he'd plop one on my desk and go 'Is this what you were aiming at?' He'd gotten so good at taking vague descriptions and making them concrete." All literacies are useful to dabble in even if you never master them. They open your eyes to the world around you. Learning to write, even clumsily, teaches us to appreciate writing done well; by learning to play an instrument, we gain a deeper appreciation of true virtuosity. The same holds with design. Try it yourself and you'll learn how hard it is to do well—and you'll appreciate more deeply what it means when someone does it at a genius level.

Even if you don't have any interest in designing objects, 3-D printers can circulate copies of thought-provoking things *other* people have designed. This can have surprising cognitive benefits. For example, complex concepts in mathematics become sensually "real" when they're transformed into physical objects. The concept of fractals becomes much more apprehensible if you hand someone a Menger Sponge, a cool-looking box that's composed of nested groups of smaller boxes. (And there are several free Menger Sponges ready for printing on Thingiverse right now.) Just as the photocopier helped us share textual and visual information, the 3-D printer helps us share *physical* information—some of which is quite historically valuable. Many museums are beginning to do 3-D scans of their artifacts, and it won't be long before schools or even individuals could print their own copy of famous historic items. If your child is wondering what a tyrannosaurus's tooth looks like, one day soon she may be able to get her library to print a copy.

The Art of Finding_

One day last fall, my wife and I were working in a café, each of us on a laptop, when she looked up.

"Hey, what's the name of that red-haired singer who plays the piano?"

I knew who she was talking about but drew a blank. I could visualize the singer sitting at her piano and playing, something I'd seen on several videos. I remembered that she'd done a piano cover of Nirvana's "Smells Like Teen Spirit."

But I couldn't attach a name. I sat there stupidly, mentally paralyzed, until suddenly I hit upon a strategy for finding the name.

"Google this," I told Emily. "'Piano cover, Smells Like Teen Spirit.'"

She bashed her keyboard and laughed: "Tori Amos!"

The exchange took no more than fifteen seconds.

Tip-of-the-tongue syndrome is an experience so common that cultures worldwide have a phrase for it. Cheyenne Indians call it *navonotootse'a,* which means "I have lost it on my tongue"; in Korean it's *hyeu kkedu-te mam-dol-da,* which has an even more gorgeous translation: "sparkling at the end of my tongue." The phenomenon generally lasts only a minute or so; your brain eventually makes the connection. But as the Tori Amos incident shows, when faced with a tip-of-the-tongue moment, many of us have begun to rely instead on the Internet to locate information on the fly. If lifelogging of the type that Gordon Bell and his peers are doing stores "episodic," or personal, memories, Internet search engines do the same for a different sort of memory: "semantic" memory, or factual knowledge about the world. When you visit Paris and have a wonderful time drinking champagne at a café, your personal experience is an episodic memory. Your ability to remember that Paris is a city and that champagne is an alcoholic beverage—that's semantic memory.

I've told that story about Tori Amos several times, and usually people chuckle because they can relate. But many of them chuckle *nervously*. They tell me that they feel uneasy about how habitually they lunge for Google (or another search engine) when trying to retrieve a piece of knowledge. I spoke to a twenty-one-year-old grad student last spring who fretted that she was getting too reliant on pinging search engines for answers: "I'm just tied to the computer," she worried. She's not alone. The design expert Don Norman once found himself cut off from the Internet while traveling to Madeira, because the roaming charges were so expensive he didn't dare launch his iPhone browser. The dependence spooked him. "I can no longer function by myself," he later recalled. "When my smartphone becomes stupid, I too become stupid." Randall Munroe, the creator of the witty Web comic *xkcd*, jokes, "When Wikipedia has a server outage, my apparent IQ drops by about thirty points."

What's the line between our own, in-brain knowledge and the sea of information around us? Does it make us smarter when we can dip in so instantly? Or dumber with every search?

This is, in a sense, the oldest technological panic out there. Writing—the original technology for externalizing information—emerged around five thousand years ago, when Mesopotamian merchants began tallying their wares using etchings on clay tablets. It emerged first as an economic tool. As with photography and the telephone and the computer, newfangled technologies for communication nearly always emerge in the world of commerce. The notion of using them for everyday, personal expression seems wasteful, risible, or debased. Then slowly it becomes merely lavish, what "wealthy people" do; then teenagers take over and the technology becomes common to the point of banality.

As writing first moved through this cycle and intruded into everyday life, it unsettled many. When I wrote about Plato's *Phaedrus* a few chapters ago, I talked about how Socrates feared that writing would kill off the Greek tradition of debate and dialectic. But he also worried it would destroy our factual knowledge. The dialogue begins when Socrates encounters his young friend Phaedrus, who has just come from hearing a speech about love by the rhetorician Lysias. Socrates asks Phaedrus to recite it, since memorizing material was, in Socrates' day, a key cognitive skill for the elites. Some intellectuals used mnemonic techniques such as memory palaces, envisioning their minds as enormous castles in which they would place snippets of text, recalling them later by mentally revisiting each room to retrieve its contents.

Phaedrus refuses to recite the speech by heart. "How," he pleads with Socrates, "can you imagine that my unpracticed memory can do justice to an elaborate work . . . ?" Instead, Phaedrus offers a sort of PowerPoint version; he'll recite a *summary* of the speech. At this point, Socrates coyly notes that Phaedrus is hiding a scroll under his cloak. It's a copy of Lysias's speech! Phaedrus was indeed trying to absorb and learn the speech, but he was relying on a more modern technique: Poring over the written word. Semantic storage was shifting from inside the brain to outside.

Socrates warned Phaedrus of the dangers of this arriviste technology. He told the story of how the Egyptian god Theuth invented writing and presented it as a gift to Thamus, the king of Egypt. "This," said Theuth, "will make the Egyptians wiser and give them better memories; it is a specific both for the memory and for the wit." Thamus disagreed. Knowledge stored was not really knowledge at all. People who became reliant on writing would lose the art of remembering anything:

> This discovery of yours will create forgetfulness in the learners' souls, because they will not use their memories; they will trust to the external written characters and not remember of themselves. The specific which you have discovered is an aid not to memory, but to reminiscence, and you give your disciples not truth, but only the semblance of truth; they will be hearers of many things and will have learned nothing; they will appear to be omniscient and will generally know nothing; they will be tiresome company, having the show of wisdom without the reality.

Thamus also observed, astutely, that one ought not to trust an inventor's hype. "O most ingenious Theuth, the parent or inventor of an art is not always the best judge of the utility or inutility of his own inventions to the users of them." That's still good advice even today. Anyone hawking their new technology won't necessarily be truthful about its implications, or even understand them.

The argument that Socrates began has never gone away. Indeed, each new improvement in publishing, archiving, and note taking has renewed its vigor. In the early Middle Ages, Chinese technicians developed advanced printing technology ahead of Gutenberg in Europe. The Chinese scholar Ye Mengde complained that written records not only reduced the art of memory ("As it was so easy for scholars to get books, their reciting them from memory also deteriorated") but increased the damage of factual errors; once enshrined in text, a false fact assumed authority and would be repeated. There was no easy way, once printed, to *amend* a book.

As printmaking processes improved in Europe and the volume of printed texts exploded, it created a very modern panic about information overload, as the scholar Ann Blair documents in *Too Much to Know*, her study of the period before and after Gutenberg. Before, when books were scarce, the literate would read and reread the few precious, rare texts to which they had access (often devotional tomes), gradually committing them to memory. But as books became plentiful—to the wealthy literate, anyway—readers shifted from "intensive" reading to our more contemporary mode of "extensive" reading: reading widely, but rarely rereading. (How often have *you* reread a book?) This shift, in turn, prompted fresh fears that nobody was retaining knowledge. "Whoever tries to retain everything will retain nothing well," wrote the twelfth-century author of the anthology *Libri deflorationum*. In the late seventeenth century, Marius D'Assigny tried to preserve the ancient art of memory palaces by publishing an instructional tome on the craft, *The Art of Memory*, which he opened with "An Address to to the Youth": "Memory is a rich and precious jewel, if polish'd, used and improved; but if suffer'd to be idle, it is as a Pearl of great Value in the hands of a slothful or ignorant Artist."

Advocates of outsourced knowledge and written records shot back. They argued that the breadth of their collections allowed them to ponder a wider array of ideas. Indeed, they disdained the ancient arts of mass memorization as a parlor trick: mnemonists might be able to retain reams of text, but this didn't guarantee they thought deeply about the material. Memory artists were "Tumblers; Buffones & Juglers" who used their skills for "ostentation," sniffed Sir Francis Bacon (who knew whereof he spoke, having been trained in memory arts as a child). The French Cartesian philosopher Nicolas Malebranche blamed rote memorization for "confusing the mind and disturbing clear ideas" while "inducing pride in the multitude of facts one had stuffed in one's head," as Blair notes. Michel de Montaigne, the famous essayist, noted that he was "a man of no retentiveness," and indeed he worked a bit like a modern blogger: sitting in his capacious library, he'd peruse a handful of unrelated books until something tickled his fancy—at which point he'd compose an essay riffing off the last few things he'd read. (Indeed, this technique may be key to Montaigne's style: finding fresh insights by linking together disparate stories and aphorisms.)

In the long run, of course, Socrates was both right and wrong. As print grew, fewer and fewer thinkers relied on mental palaces. Virtually no one today can recite long passages from memory.

But what Socrates didn't foresee was the types of complex thought that would be possible once you no longer needed to mentally store everything you'd encountered. As scholars grappled with the profusion of print, they realized that the skill of the modern age would be managing *access* to external memory. If you are going to read widely but often read books only once; if you going to tackle the ever-expanding universe of ideas by skimming and glancing as well as reading deeply; then you are going to rely on the semantic-memory version of gisting. By which I mean, you'll absorb the gist of what you read but rarely retain the specifics. Later, if you want to mull over a detail, you have to be able to *refind* a book, a passage, a quote, an article, a concept. Access becomes critical. In essence, a new and crucial intellectual skill emerged. As the purely oral world died, the world of library science was born.

Renaissance thinkers set about devising techniques to make their reading rebrowsable. This included note taking. We now regard note taking as an obvious tool for retaining the most important parts of a book, a speech, or one's own meditations. But it took centuries to codify the practice. Jeremias Drexel, a sixteenth-century Jesuit scholar of the humanities, wrote an influential manual in which he suggested maintaining separate notebooks for diverse subjects (mathematics, law, philosophy, theology). He also recommended an alphabetic index, sorting one's notes into separate, smaller notebooks. The indexing was the crucial invention, because Drexel envisioned us reading so capaciously—he himself boasted of taking notes on a mind-boggling one hundred authors a day—that there was no way normal human memory could retain it all. Indeed, like many modern readers, Drexel probably forgot not only the books he'd read, but that he'd even *made notes* on them. His indexes became a way "to enable one to retrieve notes that one might no longer remember having taken," as Blair writes. Using them, Drexel could refind material, instantly producing six pages of notes on subjects as diverse as tears, dancing, or bacchanalia.

Even as individuals worked to organize their personal stores of information, public and university librarians began to organize the shared version of this knowledge. The Gutenberg book explosion certainly increased the number of books that libraries acquired, but librarians had no agreed-upon ways to organize them. It was left to the idiosyncrasies of each. A core job of the librarian was thus simply to *find* the book each patron requested, since nobody else knew where the heck the books were. This created a bottleneck in access to books, one that grew insufferable in the nineteenth century as citizens began swarming into public venues like the British Library. "Complaints about the delays in the delivery of books to readers increased," as Matthew Battles writes in *Library: An Unquiet History*, "as did comments about the brusqueness of the staff." Some patrons were so annoyed by the glacial pace of access that they simply stole books; one was even sentenced to twelve months in prison for the crime. You can understand their frustration. The slow speed was not just a physical nuisance, but a cognitive one.

The answer? Cut the librarian out of the equation. In the late nineteenth century, Melville Dewey imposed some order. The Dewey decimal system created a taxonomy of placement: All the history books go on this shelf, all the architecture over here, and the medical stuff over there. The goal wasn't just organization, it was to dramatically quicken the pace at which books could be found. (Dewey was something of an obsessive about speed; he proselytized for shorthand and phonetic spelling, and even truncated his first name to "Melvil" and last name to "Dui" to save crucial seconds while writing). Armed with the Dewey decimal system, a reader could find books without assistance. Librarians began to reinvent their roles, transforming from mechanical grabbers-of-books to intellectual guides, helping their patrons figure out *what* to read and how to conduct complex research.

In the twentieth century, the advance of new media—photography, audio recording, telephony—inspired dreams of even huger external records. In 1910, the French information-science pioneer Paul Otlet created a sort of paper-and-ink precursor to today's remotely queryable online world: the Mundaneum, a collection of twelve million facts written on catalog cards, meticulously organized with an indexing system of Otlet's devising. For a fee, anyone could query the Mundaneum by mail or telegraph. Housed in a massive building in Belgium, it received over fifteen hundred such requests before it was finally disassembled by the occupying Nazis. But Otlet's visions of infinite access grew even grander. In 1934, he wrote an essay envisioning a great compendium of electronic documents that could be accessed via a computer in your household: a *reseau*, or "web." "Here, the workspace is no longer cluttered with any books. In their place, a screen and a telephone within reach," Otlet wrote. "Over there, in an immense edifice, are all the books and information. From there, the page to be read . . . is made to appear on the screen. The screen could be divided in half, by four, or even by ten, if multiple texts and documents had to be consulted simultaneously."

Perhaps the best-known early visionary of the Web, however, was Vannevar Bush. An American government scientist who pioneered analog computers, Bush was interested in how computers could improve human thought. In his 1945 essay "As We May Think," he envisioned a device strikingly similar to Otlet's. Like Drexel and Dewey, Bush worried that print was becoming unnavigable. "The investigator is staggered by the findings and conclusions of thousands of other workers—conclusions which he cannot find time to grasp, much less to remember, as they appear," he complained. His answer was the memex, a high-tech desk. "A memex is a device in which an individual stores all his books, records, and communications, and which is mechanized so that it may be consulted with exceeding speed and flexibility. It is an enlarged intimate supplement to his memory," he wrote. Seated at their memex, users could peruse thousands of books and articles stored on microfilm. They could write their own notes, annotate documents, and create microfilm copies of their notes to add to the record. Like Otlet, Bush envisioned users creating hyperlinks—connections between documents, the better to refind and reconsult them.

As with Drexel, Dewey, and Otlet before him, Bush argued that speed of recall was key. Without it, one's external store of facts would be useless. When he called his invention "an enlarged intimate supplement" to memory, the crucial word wasn't so much "enlarged" or "supplement"; books had long enlarged and supplemented our minds. No, it was "intimate"—the idea that the memex would be physically and cognitively proximal, in a nearly sensual fashion. That was a key to its power. Indeed, Bush suspected the comparative difficulties of using libraries is what had prevented them from being of widespread use to the public. "Even the modern great library," he wrote, "is not generally consulted; it is nibbled at by a few." To truly harness our external knowledge, we needed to bring it closer to our minds.

The history of factual memory has been fairly predictable up until now. With each innovation, we've outsourced more information, then worked to make searching more efficient. Yet somehow, the Internet age feels different. Quickly pulling up "Tori Amos" on Google seems different from looking up a bit of trivia in an encyclopedia. It's less like consulting a book than like asking someone a question, consulting a supersmart friend who lurks within our phones.

In a way, this is precisely what's going on. We're beginning to fit digital tools into another very ancient behavior: relying on social memory for facts. It turns out that historically, we store knowledge not only in the objects around us, like books. We store knowledge in the *people* around us—our friends, coworkers, and romantic partners. While we think of knowledge as an individual possession, our command of facts is often extremely collaborative.

It was in the 1980s that Harvard psychologist Daniel Wegner and his colleagues Ralph Erber and Paula Raymond first began to systematically explore this phenomenon. Wegner noticed that spouses often divide up memory tasks. The husband knows the in-laws' birthdays and where the spare lightbulbs are kept; the wife knows the bank account numbers and how to program the TiVo. If you ask the husband for his bank account number, he'll shrug. If you ask the wife for her sister-in-law's birthday, she can never remember it. Together, they know a lot. Separately, less so.

Wegner suspected this division of labor takes place because we have pretty good "metamemory." We're aware of our mental strengths and limits, and we're good at intuiting the abilities of others. Hang around a workmate or a romantic partner long enough and you begin to realize that while you're terrible at remembering your corporate meeting schedule, or current affairs in Europe, or how big a kilometer is relative to a mile, they're great at it. So you begin to subconsciously delegate the task of remembering that stuff to them, treating them like a notepad or encyclopedia. In many respects, Wegner noted, people are superior to these devices, because what we lose in accuracy we make up in speed. It takes a few minutes to walk over to the *Encyclopedia Britannica* and look up a fact, which is a bit of a hassle. (This is probably why *Britannica*'s internal studies discovered that owners opened their encyclopedias, on average, only once a year.) It's much faster to simply ask your coworker or wife, because they're right there.

Wegner called this phenomenon "transactive" memory: two heads are better than one. We share the work of remembering, Wegner argued, because it makes us collectively smarter—expanding our ability to understand the world around us.

Experiments have borne out Wegner's theory. One group of researchers studied older couples who'd been together for decades. When separated and questioned individually about the events of years ago, they'd sometimes stumble on details. But questioned together, they *could* retrieve them. How could they do this? By engaging in "cross-cuing," tossing clues back and forth until they triggered each other's memory. For example, this is how a couple remembered a show they saw on their honeymoon forty years previously:

> F: And we went to two shows, can you remember what they were called?
> M: We did. One was a musical, or were they both? I don't . . . no . . . one . . .
> F: John Hanson was in it.
> M: *Desert Song.*
> F: *Desert Song*, that's it, I couldn't remember what it was called, but yes, I knew John Hanson was in it.
> M: Yes.

In a sense, they were googling each other's memory stores. Other experiments have produced similar findings. In one, people were trained in a complex task—assembling an AM/FM radio—and tested a week later. Those who'd been trained in a group and tested with that same group performed far better than individuals who worked alone; together, they recalled more steps and made fewer mistakes. In 2009, researchers followed 209 undergraduates in a business course as they assembled into small groups to work on a semester-long project. The groups that scored highest on a test of their transactive memory—in other words, the groups where members most relied on each other to recall information—performed better than those who didn't use transactive memory. And a similar study found that the higher-scoring groups didn't just recall facts better; they *analyzed* problems more deeply, too, developing a better grasp of underlying principles.

We don't remember in isolation—and that's a good thing. "Quite simply, we seem to record as much outside our minds as within them," as Wegner has written. When information is located inside someone you trust, it's a mind expander: "Couples who are able to remember things transactively offer their constituent individuals storage for and access to a far wider array of information than they would otherwise command."

These are—as Wegner describes it in a lovely phrase—"the thinking processes of the intimate dyad."

Transactive memory helps explain how we're evolving in a world of on-tap information.

Five years ago, a young grad student of Wegner's, Betsy Sparrow, was watching a movie with her husband. Stuck on the name of one of the actors, she googled it. That made her wonder: Maybe people were using search engines, and digital retrieval, as transactive memory. Had search tools like Google become so omnipresent—so "intimate," as Vannevar Bush foresaw—that they rivaled the ease of asking your spouse?

To test this, Sparrow ran an experiment. She took a handful of students, gave them sentences of trivia, and had them type the sentences into a computer—factoids like "An ostrich's eye is bigger than its brain" and "The space shuttle *Columbia* disintegrated during re-entry over Texas in Feb. 2003." With some facts, the students were explicitly told the information wouldn't be saved; with others, the screen would tell them that the fact had been saved in one of five blandly named folders, such as FACTS, ITEMS, or POINTS. When Sparrow tested the students, the people who knew the computer had saved the information were less likely to personally recall the info than the ones who were told the trivia wouldn't be saved. In other words, if we know a digital tool is going to remember a fact, we're slightly less likely to remember it ourselves.

We are, however, confident of where we can refind it. When Sparrow asked the students simply to recall *whether* a fact had been saved or erased, they were better at recalling the instances in which a fact had been stored in a folder. As she wrote in a *Science* paper, "believing that one won't have access to the information in the future enhances memory for the information itself, whereas believing the information was saved externally enhances memory for the fact that the information could be accessed." Each situation strengthens a different type of memory.

To explore this phenomenon further, Sparrow did one more experiment. She had the students type in the factoids, and after each one they'd be told the name of the folder it had been saved in. When she tested their recall, they were twice as likely to correctly name the folder where it had been saved—despite the deliberately bland folder names—than to remember the fact itself. On a subconscious level, the students clearly recognized that if they weren't going to commit a detail to memory, they still needed to remember where they could find it.

For Sparrow, this suggested her theory was right: When we're surrounded by computer memory, we treat it the same way we treat other people. Google—or your smartphone—becomes like an insanely knowledgeable companion. You trust it enough to rely on as you would a spouse or nearby colleague.

"Just as we learn through transactive memory who knows what in our families and offices, we are learning what the computer 'knows' and when we should attend to where we have stored information in our computer-based memories," Sparrow wrote. "We are becoming symbiotic with our computer tools."

You could say this is precisely what we most fear: Our mental capacity is shrinking! But Sparrow isn't so sure this is anything particularly new or particularly bad. After all, she tells me, we've been using transactive memory for millennia with other humans. In everyday life, we are only rarely isolated, and for good reason. For many thinking tasks, we're dumber and less cognitively nimble if we're not around other people. Not only has transactive memory not hurt us, it's allowed us to perform at higher levels, accomplishing acts of reasoning that are impossible for us alone. It wasn't until recently that computer memory became fast enough to be consulted on the fly, but once it did—with search engines boasting that they return results in tenths of a second—our transactive habits adapted.

In fact, as transactive partners, machines have several advantages over human ones. For one, the fidelity of their recall doesn't hazily decay, the way ours does. As we found earlier when talking about autobiographical memories, we humans are good at preserving the gist of something but bad at the details. The same thing is true with factual knowledge: We easily retain the general contours of a piece of knowledge, but we're terrible at the small print. In 1990, psychologist Walter Kintsch documented this in a fascinating experiment. He showed people a bunch of sentences, then tested them on their recall. He found that forty minutes after reading the material, the subjects could still remember the precise wording of the sentences. But this performance quickly degraded, such that four days later they recalled almost nothing of the word-by-word specifics. In contrast, their ability to remember the "situation model"—the events being described in general—didn't decline at all. Sure, they'd lost the details, but they kept the meaning. They had taken what the sentence was *about* and incorporated it into their general understanding of the world, or what psychologists would call their schemas.

This phenomenon is familiar to everyone who reads the daily newspaper. On Monday you read a story about drone strikes in the Waziristan area of Pakistan and how the local Pashtun population reacts angrily. A few days later, you won't recall the specifics. But you will probably have incorporated the knowledge into one of your schemas for understanding the world: "Drone strikes are seriously angering many Pakistanis." If you're in a bar a few weeks later arguing about this subject, you'll lunge for a search engine or Wikipedia to remind you of the specifics—the *precise* evidence that supports what you *generally* know.

In some ways, machines make for better transactive memory buddies than humans. They know more, but they're not awkward about pushing it in our faces. When you search the Web, you get your answer—but you also get much more. Consider this: If I'm trying to remember what part of Pakistan has experienced many U.S. drone strikes and I ask a colleague who follows foreign affairs, he'll tell me "Waziristan." But when I queried this once on the Internet, I got the Wikipedia page on "Drone attacks in Pakistan." A chart caught my eye showing the astonishing increase of drone attacks (from 1 a year to 122 a year); then I glanced down to read a précis of studies on how Waziristan residents feel about being bombed. (One report suggested they weren't as opposed as I'd expected, because many hated the Taliban, too.) Obviously, I was procrastinating. But I was also learning more, reinforcing my schematic understanding of Pakistan.

Now imagine if my colleague behaved like a search engine—if, upon being queried, he delivered a five-minute lecture on Waziristan. Odds are I'd have brusquely cut him off. "Dude. Seriously! I have to get back to work." When humans lecture at us, it's boorish. When machines do it, it's fine. We're in control, so we can tolerate and even enjoy it. You likely experience this effect each day, whenever you use digital tools to remind yourself of the lyrics to a popular song, or who won the World Series a decade ago, or what the weather's like in Berlin. You get the answer—but every once in a while, whoops, you get drawn deeper into the subject and wind up surfing for a few minutes (or an entire hour, if you vanish into a Wikipedia black hole). We're seduced into learning more, almost against our will. And there are a lot of opportunities for these encounters. Though we may assume search engines are used to answer questions, some research has found that in reality up to 40 percent of all queries are acts of *remembering*. We're trying to refresh the details of something we've previously encountered.

The real challenge of using machines for transactive memory lies in the inscrutability of their mechanics. Transactive memory works best when you have a sense of how your partners' minds work— where they're strong, where they're weak, where their biases lie. I can judge that for people close to me. But it's harder with digital tools, particularly search engines. You can certainly learn how they work and develop a mental model of Google's biases. (For example: Google appears to uprank sites that grow in popularity slowly and regularly, because those sites are less likely to be spam.) But search companies are for-profit firms. They guard their algorithms like crown jewels. This makes them different from previous forms of outboard memory. A public library keeps no *intentional* secrets about its mechanisms; a search engine keeps many. On top of this inscrutability, it's hard to know what to trust in a world of self-publishing. To rely on networked digital knowledge, you need to look with skeptical eyes. It's a skill that should be taught with the same urgency we devote to teaching math and writing.

Learning the biases of digital tools can also help us avoid a pitfall of transactive memory called collaborative inhibition. This happens when two people have methods of remembering information that actively *clash* with one another. In this situation, they can actually remember less together than they do separately. I suspect these clashes can also occur with machines. It explains why we often have such strong opinions about what type of transactive-memory tools we use. I've seen professionals nearly come to blows over which tool is "best" for storage and recall: Apps like Evernote? Plain-text documents? Complex databases like Devonthink? Saving your notes in a Gmail account and querying it? Each person has gravitated to a tool that they understand well and that fits their cognitive style.

The subtler risk of living in our Google-drenched world may not be how it affects our factual knowledge. It's how it affects our creativity.

Sure, social knowledge is great. But many big strikes of creative insight really do require *Thinker*-style interiority, in which we quietly mull over material that we have deeply internalized. In his essay "Mathematical Creation," Henri Poincaré described the way this sort of breakthrough happens. Poincaré had been working for two weeks at his desk, trying to prove that a set of functions couldn't exist. One day he overcaffeinated himself, and while he lay in bed trying to sleep, the ideas stirred in his head. A breakthrough emerged: By morning he had realized that "Fuchsian functions"—as he eventually called them—really *did* exist. Poincaré went back to work at his desk but found himself again unable to make headway. So he took another break, this time leaving on a geological excursion. Again, he wasn't actively working on the problem—but some deep part of his mind was clearly still turning it over, because as he stepped onto a bus, a solution popped into his head. Back to his desk he went for more work, and again he hit a roadblock. This time Poincaré was called away for compulsory military service, and again while he was on the road, an answer popped into his head unbidden—another bolt from the blue, a gift from his subconscious mind.

These eureka moments are familiar to all of us; they're why we take a shower or go for a walk when we're stuck on a problem. But this technique works only if we've actually got a lot of knowledge about the problem stored in our brains through long study and focus. As Poincaré's experience illustrates, you can't come to a moment of creative insight if you haven't got any mental fuel. You can't be googling the info; it's got to be inside you. When I was in college studying Western mythology, literary critic Northrop Frye told our class that he'd been able to write his masterpiece *The Great Code*—a synthesis of the structure of the Bible and its relation to Western literature—because he'd read the Bible so many times he'd essentially memorized it; the same goes for Western poetry, a lot of which he'd also committed to memory. Frye could rotate the Bible and the Western canon in his mind like a crystal lattice, which is how he was able to spot their shared mythic patterns. In 1980, essayist Clara Claiborne Park wrote a lament for the demise of memorization. (At the time, the villains were pocket calculators and TelePrompTers.) Park feared that creativity would suffer. The

Greek muses, she pointed out, were the daughters of Mnemosyne, the personification of memory.

This is a genuine concern. Could we reach a dangerous area, where we've put so many details out of mind that creative analysis becomes harder, or impossible?

Right now, I doubt it. Evidence suggests that when it comes to knowledge we're interested in—anything that truly excites us and has meaning—we don't turn off our memory. Certainly, we outsource when the details are dull, as we now do with phone numbers. These are inherently meaningless strings of information, which offer little purchase on the mind. (Names, interestingly, have the same problem: as the philosopher John Stuart Mill complained, they're not "connotative." "Clive" doesn't inherently mean anything, not in the way that "fire hydrant" does.) It makes sense that our transactive brains would hand this stuff off to machines.

But when information engages us—when we really care about a subject—the evidence suggests we don't turn off our memory at all. In one fascinating 1979 experiment, scientists gave a detailed description of a fictitious half inning of baseball to two groups: one composed of avid baseball fans, the other of people who didn't know the game well. When asked later to recall what they'd read, the baseball fans had "significantly greater" recall than the nonfans. Because the former cared deeply about baseball, they fit the details into their schema of how the game works. The nonfans had no such mental model, so the details didn't stick. A similar study found that map experts retained far more details from a contour map than nonexperts. The more you know about a field, the more easily you can absorb facts about it.

But this doesn't mean you have a better memory for *all* information. Although we talk about someone's having a mind like a steel trap, memory for facts is quite specific to our obsessions, as this research shows. Outside your field of specialty, your ability to absorb facts isn't going to be above average. The "absent-minded professor" has a capacious recall for academic minutiae in his field but no idea where he parked his car. A Pokemon-loving kid remembers every monster but zones out in history class. Feminists, of course, have long noted that "great men" use their wives for domestic transactive memory: They can focus on work because someone else is doing the drudge labor of managing the socks.

If passion is the force that powers our focus and our memory, the trick is how to seduce people to be passionate about a broad array of things. Sure, it's great that baseball and Pokemon fans can gobble up stats for their respective crazes, but wouldn't it be better for society if they also had that level of fascination for, say, civic life? In an ideal world, we'd all fit the Renaissance model—we'd be curious about everything, filled with diverse knowledge and thus absorbing all current events and culture like sponges. But this battle is age-old, because it's ultimately not just technological. It's cultural and moral and spiritual; "getting young people to care about the hard stuff" is a struggle that goes back centuries and requires constant societal arguments and work. It's not that our media and technological environment don't matter, of course. But the vintage of this problem indicates that the solution isn't merely in the media environment either.

When I first began researching the effect of digital tools on factual knowledge, I interviewed several leading memory experts, hoping they could explain where we're headed. Alas, they couldn't. With admirable candor, they all admitted to the limits of their knowledge. If you're a memory scientist, you often get the "is my iPhone destroying my memory?" question at cocktail parties, but you can't easily answer it. Nobody has yet studied the long-term effects of relying on external, intimate memory tools. "That's a hard study to do to get a meaningful answer from," as Daniel Schacter, head of Harvard's Schacter Memory Lab, told me. How would you get a control group? Raise a thousand kids to the age of twenty-five without letting them have access to computers? Or libraries? Or newspapers, or even *paper*?

Schacter suspects we'll evolve as we always have, using the same mental flexibility that let us travel from papyrus scroll to Post-it note. "The question of whether it's somehow going to reduce memory overall is an intriguing one. But I think there's a clear benefit from using external aids and it's unknown whether there are any costs," he said. From Socrates onward, we lose old cognitive skills as we gain new ones, but we've benefited from the trade-off.

For my money, there's a far more immediate danger to the quality of our in-brain memory: that old op-ed page demon, distraction.

If you want to internalize a piece of knowledge, you've got to linger over it. You can't flit back and forth; you have to focus for a reasonable amount of time, with mental peace. But today's digital environment rarely leaves you any such peace. Indeed, most of today's tools are designed to constantly interrupt us—to beep when new e-mail arrives, a friend checks in, or a photo gets posted. Each app is fighting for you to spend more time with it. That's partly because of commercial dictates, particularly with social media services supported by advertising, where the companies make money only if they can get people hooked on using their product. And it's also partly because the engineers writing the software often operate from the false assumption that humans function the way machines do, happily shifting from task to task without strain.

But humans don't work that way. Constantly switching between tasks is ruinous to our attention and focus. When informatics professor Gloria Mark studied office employees for one thousand hours, she found that they could concentrate for only eleven minutes at a time on a project before being interrupted or switching to another task—and once they'd been interrupted, it took an average of twenty-five minutes to return to their original work. Other research has confirmed that rapid task switching makes it harder to manage our attention and to retain what we read. In one experiment, students who watched lectures while sending text messages did roughly 19 percent worse on a test than nontexting students. We develop what Nicholas Carr dubbed the "juggler's brain," a mind that can't learn things because it doesn't stand still long enough.

Of course, task switching isn't always bad. Certain kinds of work actually require and reward it. One survey of 197 businesses found that those with top management who were heavy multitaskers got 130 percent better financial results than companies that worked in a more "monochronic" fashion. Why? Because the multitasking employees were more plugged in to their peers, able to vacuum up information and make decisions more quickly. It made them more efficient. The problem is that efficiency isn't always the goal with cognition. If you want to deeply absorb knowledge, you often want to work *inefficiently*—lingering and puzzling and letting ideas sink in. We need to do both, but our current digital environment is mostly designed to favor multitasking.

Is there a technological fix for this technological conundrum? Could better design reduce interruptions? A few promising new interfaces have emerged, including the crop of "read it later" services (like Instapaper or Pocket) that let people take long-form writing from the Web—which is hard to absorb on the fly—and push them to tools like Kindles for a better reading experience; evidence suggests we comprehend as much when we read on these devices as when we read on paper. There are also apps like Freedom that shut down connectivity entirely.

But realistically, I suspect there's no killer app to end distraction. The downsides of being highly networked are constitutionally tied to the benefits. The only way we can reduce the negative side effects is by changing our relationship to the digital environment, both as a society and as individuals.

On the societal level, we could start by trying to rein in the incessant demands of employers. White-collar distractedness isn't an inevitable by-product of technology. Design plays a role, but the crisis is arguably driven just as much by how employers use today's technological efficiency and portability to squeeze ever more labor out of their workers—by, say, demanding they check their e-mail 24-7. If you change this corporate behavior, distractability changes, too. When Gloria Mark ran an experiment in which thirteen office workers went five days without using e-mail at all, there were tremendous positive effects: "task switching" dropped dramatically (they flipped between windows about half as often as before), their stress went down (as measured by their heart rate variation), and they reported greater ability to focus on single projects. Obviously, going cold turkey on e-mail isn't viable for many workers (though as I'll explore later in this book, some have done so). The problem of technological overwork requires a collective political response. Religious traditions and trade unions understood this, back when they lobbied for the Sabbath and weekends off for workers. Digital sabbaths are crucial for cognition and for the spirit.

But there are things any individual can do, too, including learning to better control your flighty brain. New-agey though this may sound, one surprisingly useful approach is doing a bit of meditation. As the author Maggie Jackson writes, the essential ethic of meditation is cultivating mindfulness: paying attention to your attention. Practice meditating and you begin to gain "an overarching ability to watch and understand your own mind," as Jackson puts it.

A friend of mine, the playwright Julia May Jonas, once described the power that mindfulness gave her. She'd started doing some basic exercises in which she'd focus on her breathing. When her mind drifted—as it inevitably would—she'd notice whatever subject she'd drifted off to (anxieties about work, lyrics to a song, "Did I leave the oven on?"). Then she'd return to her breathing. This cycle would repeat: her mind would drift, and each time she'd bring it back.

Soon, she noticed something delightful: Her mind started drifting less while she was working at her computer, too. She could focus on tasks for longer. When an inevitable distraction came along, she was much better at quickly returning to her original task, because she could notice the distraction and mentally track it, rather than simply responding like Pavlov's dog. Better yet, she became better at resisting "internal" distractions, her mind's sudden desire to interrupt herself, to go check Twitter or Tumblr for a pellet of novelty. Everything, she told me, got easier and calmer in some very subtle way.

We're getting closer to the machine all the time; our transactive partners now ride in our pockets. But what will happen as they move even closer to our minds?

If you want a glimpse of that, say hello to Thad Starner.

Starner is a forty-three-year-old computer science professor at the Georgia Institute of Technology who also works for Google. But he's best known as one of the few people on the planet with years of experience using a wearable computer. The guts of the computer are the size of a small softcover book, strapped to his torso in what amounts to a high-tech man purse. He types into it using a Twiddler, an egg-sized device that lets him write with one hand. And what's most prominent is the screen—a tiny LCD clipped to his glasses, jutting out just in front of his left eyeball. While you or I have to pull out a phone to look up a fact, he's got a screen floating in space before him. You might have seen pictures of Google Glass, a wearable computer the company intends to release in 2014. Starner's helping Google build it, in part because of his long experience: He's been wearing his for two decades.

"This is about creating a higher level of intellect—an augmented intellect," he tells me when I meet him. Starner has a cheery, surfer-like handsomeness, but the tiny black protuberance jutting out of his glasses is at first pretty jarring. "Actually, this thing gets me into any party I want to go to," he laughs. Ten years ago, onlookers assumed Starner's wearable was a medical device that mitigated some defect in his vision, and they gave him a wide berth. Now Swedish investment bankers beg him to tell them where they can buy one. Like the small group of other long-term users of wearables world-wide—there are only about twenty others who've done this for years—Starner built his rig by hand.

Starner first got the idea in 1989 in an undergraduate class at MIT. He was frustrated by a paradox in note taking: Whenever he wrote a note, he stopped paying attention to the professor and risked missing out on new information. But if he *didn't* take notes? He'd have nothing to study from. He wanted to split the difference—to have a computer that was so integrated into his body and his field of vision that he could keep looking at the lecturer while he typed.

Starner knew about similar prototypes. The first wearable was cocreated in 1960 by Claude Shannon, the founder of information theory, for a purpose both whimsical and mathematically ambitious: He wanted to beat the Las Vegas roulette wheels. Using sensors built into his shoes, Shannon would "type" information about how the wheel had been spun, and a circuit in his pocket would send beeps to an earpiece telling him how to bet. (The question of whether wearable computing constitutes amplified smarts or blatant cheating hinges, I suppose, on whether you identify with the casino owner or the players.) In a few months, Starner had cobbled together his own wearable. It was as heavy as a textbook, but it worked.

He became a cyborgian Jeremias Drexel. Because the keyboard was always in his hand, Starner would write down stray thoughts, then bring them up in conversation later on. "I'm basically an expert on anything I've ever heard anyone talk about, because I can pull it up and reread it at a moment's notice," he says. He opens his laptop to show me what his screen looks like, picking a subject at random—*artificial nose technology*. He's got dozens of lines of notes, beginning when he first attended a lecture on artificial smell sensors at MIT in the 1990s and going up to last year. ("Smell is mostly learned" is one note; "resistors made by conductive inks" is another. "Wound healing"—that puzzles me. What's that? "As wounds heal, the bacteria outgases differently," he says, "so you can make bandages that smell the bacteria and report on the status of the wound.") These notes seriously enrich the experience of talking to Starner, because he's adept at picking up subjects you discussed a year ago, then musing over his thoughts since. In two decades, he's jotted down two million words of notes.

Starner's wearable is also connected online, producing moments of seeming omniscience that can be startling. While we're having coffee at a Starbucks, I casually mention Timehop, that service that replays your status updates from a year ago. "Right," he nods, then recites their corporate slogan: "The 'time capsule of you.'" He'd pulled up the Web site while I was talking about it. Later, when I mention a story I'd written a decade ago, he casually drops in a few precise details about it, too.

Starner has done even weirder tricks, including once swapping his entire life's worth of notes with a friend who was also using a wearable. They'd occasionally find themselves answering questions by finding notes the other had made years earlier—answering questions by consulting someone *else's* memory, as it were. And he's used it to do public thinking on the fly. While testifying at a National Academy of Sciences hearing into the impact of wearables, he had some of his grad students listening in to the event and offering him tidbits of information to help answer tough questions. "I'll have people backing me up. I seem more smart than I am!" he jokes.

Part of what makes Starner's wearable so useful is that it's so fast to consult. Despite the idea that we Google "everything" on our smartphones, in reality it's still a tiny hassle to pull it out of our pockets, so we do it only when we really feel a hankering, as when settling an argument or checking a train schedule. The same thing goes with taking notes. We almost never do it, because it's awkward, not just physically but socially; you don't want to interrupt the flow of a conversation by jotting something down. By putting a screen right in his eyesight and a keyboard in his hand, Starner has effectively moved the mobile-phone factoid out of his pocket—where he'll rarely consult it—and into his peripheral vision, where he almost always will.

I tell him many people would feel too self-conscious to wear a screen poking out in front of their face like that, or to hold a one-handed keyboard. He's not so sure. He points out that Bluetooth headsets used to look odd, too. (In fact, one wearable user I interviewed is dyslexic, so he has his computer "read" information to him quietly in his ear, like James Bond's "Q.") And it's also true that among my friends in the 1990s, the mere possession of a mobile phone marked you as a self-important boor. *That* judgment certainly vanished. There's already a joking insult for the Google developers who've been spotted wearing Glass—"Glassholes"—so it's possible the same pattern will emerge with wearables: outright mockery, slow adoption, then Beyoncé will be seen wearing one onstage and the craze will begin. Starner also suspects other tools will bridge the gap, like wristwatch computers. (The biggest problem with Glass will likely stem from something else, though: the camera pointing out of it. Critics have already noted how unsettling it might feel to be near someone with a head-mounted camera. In fact, some early wearable pioneers didn't put cameras on their rigs for that very reason, and Starner didn't have one on his wearable either.)

Doesn't he get distracted with a screen in his field of vision? I couldn't imagine how he'd be able to pay attention to anything. (Some people who've hung out with Starner say they can notice his eyes occasionally darting over to the screen, though I couldn't.) But Starner doesn't come off as a high-tech ADD head case. Compared to some New Yorkers I see around me in everyday life, witlessly absorbed by their iPhones as they walk down the street, he's very attentive and polite.

That's because, he says, he obeys strict social protocols. He uses his wearable only to look up information that augments a conversation he's having. If he's talking with someone about the Boston Red Sox, he might pull up statistics to sprinkle in, but he's not secretly perusing cute-cat videos. "If you're finding information that supports your conversation, that's making it richer; it's multi*plexing*, not multitasking," he argues. He waits for natural pauses in conversation to jot down notes or to consult them. And he has an absolute ban on checking e-mail when face-to-face.

"If you check e-mail while you talk to someone you lose forty IQ points," he says. "People can't multitask. It's not possible. I think attention is a big, big issue. People are addicted to their Crackberries. You've got to make the systems so that they help people pay attention to the world in front of them."

I pause to absorb this directive, coming from a guy with a computer he sticks onto his face. You could argue either that he's deluding himself or that he's found a way to benefit from instant knowledge without losing his soul. I would make the second argument: Starner's whole point is that his wearable allows him to keep in attentive eye contact with the people and the world around him. Indeed, I noticed this difference between how he and I used our tools. One time when I visited him in Atlanta, we were walking across the street when I got caught up in checking a map on my iPhone to locate my hotel. Oblivious, I nearly walked into a huge, water-filled pothole; it was Starner who noticed it and warned me.

Curious to see what it's like to have a computer in your eye, I tried out Starner's eyepiece. When I held it up, I saw what looked like a crisp screen floating at the right-hand side of my right eye. I could easily read Starner's on-screen notes, which were all from his and my conversations. It felt odd, but it also felt like something I could get used to.

Perhaps most interesting is his distinction between the psychology of finding something and *re*finding it. If he's alone and doing research, he'll use his wearable to google documents, just like any of us sitting at a computer. But when he's talking to someone, he'll mostly just ping his notes; he searches online much less often. That's because his notes are his personal semantic stores; he's recuing facts, refreshing the details of what he already generally knows. That process is fast and doesn't distract. But trying to imbibe a *new* fact requires focus and attention, so he avoids doing it while in conversation.

Starner doesn't think his use of on-tap recall has eroded his own memory. "It's actually the opposite," he argues. His recall of arcana is strengthened by repetition. "If you pull up the same fact seven or eight times, eventually you've been reencountering it so often that you wind up remembering it unaided," he says. That is indeed what technological pioneers envisioned in their dreamy, visionary manifestos. When Vannevar Bush outlined the memex, he argued (as Drexel had centuries earlier) that a pocket library is useful only if you visit it again and again. It's those refindings and remusings that spark meaning and insight. "A record, if it is to be useful," Bush wrote, ". . . must be continuously extended, it must be stored, and above all it must be consulted."

———————

Research suggests that Starner is onto something: Used the right way, digital memories amplify what people retain in their brains. For example, neuroscientists who work with Microsoft in the United Kingdom have taken elderly patients with severe memory loss resulting from Alzheimer's and encephalitis and outfitted them with SenseCams. Normally, the patients would retain memories for only a few days, quickly forgetting most of what they'd done. But the researchers got the patients to spend a few minutes each night looking over the visual record of their day. Their recall improved, sometimes dramatically. Some retained memories for weeks and months—one for an astounding ten months. Alan Smeaton, another computer scientist who's worked with SenseCams, noted the same effect in himself and his students. They'd wear SenseCams all day long; then, while downloading the thousands of pictures each night, they would inadvertently relive their day, flipping through images as they loaded onto the hard drive. "I'd be reminded of all these things I'd forgotten," Smeaton tells me. "I'd see myself stop to talk to a colleague in the hallway and realize I'd totally forgotten the interesting conversation we'd had and that I'd promised to send him a paper I'd read."

Machines can also remind us of facts precisely when we need reminding. If you'll recall the Ebbinghaus curve of forgetting from the second chapter, Ebbinghaus found that we forget things in a predictable pattern: More than half our facts are gone in an hour, about two thirds are gone within a day, and within a month we're down to about 20 percent. Ebbinghaus and his followers theorized that this process could work in reverse. If you *reviewed* a fact one day after you first encountered it, you'd fight the curve of loss. This process is called "spaced repetition," and experiments and anecdotes suggest it can work. It explains why students who cram for a test never retain much; the material dissolves because they never repeat it. But though spaced repetition is clever and effective, it has never caught on widely, because ironically, the technique relies on our frail human memories. How would you remember to review something the next day? Then a few days later, a week, and three months?

Machines, however, are superb at following these rote schedules. In the last decade, software programmers began selling tools intended to let you feed in facts, which the computer then reminds you to review on a reverse Ebbinghaus curve. Use of this software has remained a subculture, mostly by people seeking to learn a foreign language, though devout adherents use it to retain everything from recipes to poetry. (One user I interviewed complained, "I wasted my whole time in high school because I didn't have this software to help me learn and remember stuff.") And you may get a chance to try it yourself soon, because the concept is gradually leaking into the design of more mainstream digital technology.

For example, I was surprised a year ago to discover that Amazon had designed a tool called Daily Review, which would take Kindle e-book notes and present them for review, in an Ebbinghausian fashion. I'd recently marked up several e-books—including *Moby-Dick*, a short tome on Chinese dissidents, a Garry Kasparov book on chess, the sci-fi novel *Ready Player One*—and decided to give the software a try. Each day, the software offered me a handful of passages. At first, I doubted there was much value to this, however pleasant it was to reencounter the snippets of text. They reminded me of how much I'd enjoyed the book, and also how much I'd forgotten. I realized that with each book, I'd kept the gist of the plot or argument—Moby-Dick *is about this total freak who chases a whale to his watery grave*—but specific details? They were sparse. Quite often my mind was bare as a birch in winter.

Yet as the weeks wore on and the passages repeated, they became more and more familiar, less like a happy surprise (oh, I'd forgotten that!) than like an echo. Pretty soon, I could summon the demise of the *Pequod*—"the great shroud of the sea rolled on as it rolled five thousand years ago"—with startling accuracy. The machine had improved my mind.

I told Starner about the experience. We were having lunch near Google's office in California on a sunny day, and when he bicycled up he wasn't wearing his device. The cable leading up to his eyeglass-mounted screen had broken. "It's the most expensive part of the entire system, believe it or not," he said, pulling it out and exposing the bundle of colored wires inside like a cortex. We talked for an hour, during which he regaled me with stories and stray bits of research.

He did pretty well without his outboard brain, I told him. He laughed and agreed.

The Puzzle-Hungry World_

How did video games get so complex?

Let's take, for example, the hit 2011 game *The Elder Scrolls V: Skyrim*. It's a Tolkien-like role-playing game, where you play as an adventurer within a sprawling medieval world. You fight scores of foes, from mace-wielding giants to fire-breathing dragons, and face hundreds of quests. Those quests increase in difficulty, so you're constantly "leveling up" your character. Leveling up, in turn, requires pondering which of the eighteen possible skills (such as archery, lock picking, and alchemy) you ought to be mastering, which type of armor and weapon to buy and wield (there are thousands of combinations), and which bonus powers—"perks"—to purchase with your experience points. (There are more than 240 different perks.)

Even longtime gamers were stunned by *Skyrim*'s byzantine folds. "I've poured over a hundred hours into the game so far and still have no idea how many more quests may be lurking out there," a game reviewer for *Forbes* magazine marveled.

Of course, video games used to be drop-dead simple. The first hit arcade game—1972's *Pong*, a digital version of Ping-Pong—required only a few lines of instruction: "Deposit quarter/Ball will serve automatically/Avoid missing ball for high score." As *Pong* innovator Nolan Bushnell once noted, it was a game "simple enough for a drunk to play." The other early arcade greats were similarly stripped down. In *Space Invaders*, you shot the aliens as they thudded slowly down the screen. In *Pac-Man*, you avoided the ghosts and ate the dots. In *Asteroids*, you shot asteroids and UFOs and avoided getting hit by anything. Teenagers around the world crammed into arcades to try to master these simple games.

Then something happened: The teenagers began sharing information.

If you were hanging out in the arcades in the 1980s—as I was, in Toronto as a teenager—you spent a lot of time not merely playing the games but talking about them. While one kid played, a cluster of others would gather around the machine. You'd share bits of information about how best to beat them, learning tricks from your peers. Together, you began to discover odd quirks in the machines' behavior. For example, elite *Pac-Man* players could figure out the dot-eating pattern that would allow you play for hours on a single quarter. There was also an exciting flaw in the game *Galaga*— the "no fire" bug. On the first stage, you could shoot all the aliens except for the bottom-left one. Then you let that one attack you and reattack while you dodged it, for fifteen minutes. If you did this, it would stop stop firing, and once you killed it, none of the aliens on any later levels would fire a single shot, making it possible to rack up a monstrously high score. Many games had similar bugs, little loopholes in the programming that created weaknesses a clever player could exploit.

If you were playing alone, discovering those bugs might take you hundreds of hours. But you weren't playing alone. Arcades were social networks, and players were avid traders of information.

As a result, those early games never stood a chance. They gave up their few mysteries in a matter of weeks. Collectively, the network of gamers was too smart.

Pretty soon, designers noticed this collective intelligence of gamers—and began to respond. They made the games intentionally more complicated, filled with "Easter eggs," little secrets to be uncovered. The first famous one was programmed into the Atari game *Adventure*. The designer, Warren Robinett, placed a tiny dot in one dungeon; if you somehow discovered it and picked it up, it'd open a gateway to a secret room that spelled out Robinett's name. Over at Nintendo, designer Shigeru Miyamoto began including hidden zones in *Super Mario* games, where players could scoop up dozens of golden coins.

Then the Internet came along in the 1990s, and the collaborative smarts of gamers exploded. Gamers used Usenet discussion boards to pool their knowledge, creating lists of strategies. They became astonishingly fast at sussing out arcana and documenting it. When Sega released *Virtua Fighter 3* in 1996, the game included eleven characters, with hundreds of hidden fighting moves. Barely a few months after the game had been released, I could find comprehensive lists of every character's moves online, painstakingly assembled by gamers around the world. (Want to get the character Wolf Hawkfield to do a "double punch-elbow-double-arm suplex throw"? Just use this button combination: "Punch, punch, tap-forward, punch, tap-down, tap-backward, punch, kick, guard.")

If a game didn't have this sort of complexity, gamers got bored and moved on. The fun in a game isn't in having mastered it. It's in the process *of* mastery—of figuring out the invisible dynamics behind how the thing works, revealing its secrets. By the early 2000s, hidden material in a game was no longer window dressing. It had been transformed into one of the central pleasures. Designers were producing worlds jammed with increasingly obtuse puzzles, storylines, distant areas, and unspoken secrets, because gamers would complain if there wasn't enough to figure out. And because they were working together collaboratively, they could figure almost anything out. In a sense, the collective smarts of players produced a cognitive arms race—with designers forced to produce ever more immense and complex imaginary universes.

This is precisely why something as insanely convoluted as *Skyrim* can exist. Individual gamers don't find such a game daunting, because they're not playing it as individuals. Within hours of the game's release in November 2011, fans set about documenting its every nuance. Soon there was a wiki with 15,789 pages, including dozens of walk-throughs—descriptions of quests—complete with maps and screenshots, and boards teeming with tips and strategies on managing your inventory. ("You can get an invincible companion dog to fight for you with another human companion at the same time by speaking to Lod at Falkreath to find his dog Barbas.")

To be sure, not all gamers use these collective documents. It's fun to solve problems on your own, so research shows most players use crowdsourced documents sparingly, such as when they're stuck or pressed for time. But when it comes to the biggest, most sprawling games—like *World of Warcraft*—the number of players who dip into crowd wisdom is huge: According to one estimate in 2008, about 50 percent of English-speaking *World of Warcraft* players were using the game's wiki every month. Top players form "guilds" that play together, cooperating to tackle the most difficult monsters.

This collective intelligence shows up even in groups of gamers as small as two. Three years ago, the designer and developer Matt T. Wood began working on the sequel to the hit game *Portal*. In the original *Portal*, gamers attempted to escape puzzling rooms by using a "portal gun" that would let them teleport from one part of a room to another. The game was praised for posing some very tricky mind-and-physics-bending challenges. The sequel, *Portal 2*, included a co-op mode in which two people, each wielding his own portal gun, cooperated as they tried to jointly escape the rooms. Wood and his co-op teammates created puzzles that they thought were suitably difficult for a pair of gamers to solve together. But when they tested it out in real life, the co-op players breezed through much faster than Wood had expected. How could they be so good?

The two players would often keep an audio channel open, brainstorming and building off each other's ideas—so "that puzzle that would have taken you fifteen minutes to solve solo only takes you two minutes to solve with someone else working with you," as Wood tells me in an e-mail. Two heads weren't just better than one; they were exponentially so. "We didn't anticipate how much harder the players would allow us to make it," Wood notes. He and his teammates went back to the drawing board, crafting rooms that were far more challenging.

In a sense, the video-game world was the first major industry to encounter the cognitive power of a highly connected audience. And what did it learn? That a vast community of networked people isn't just smart—it's restless and *hungry* for complex problems. How can we tap that resource? What types of problems is collective wisdom good at solving?

The reason millions of people collaborate on playing video games is that millions of people can now collaborate on anything. As Clay Shirky wrote in his book *Here Comes Everybody*, society has always had latent groups—collections of people all obsessed with the same thing and wishing they could work together on it. This is what the theory of multiples would predict, of course: If you're fascinated by subject X, no matter how obscure and idiosyncratic, a thousand people are out there with the same fascination.

But for most of history, people couldn't engage in mass collaboration. It was too expensive. To organize a widespread group around a task in the pre-Internet period, you needed a central office, staff devoted to coordinating efforts, expensive forms of long-distance communication (telegraphs, phone lines, trains), somebody to buy pencils and paper clips and to manage inventory. These are known as transaction costs, and they're huge. But there was no way around them. As Shirky points out, following the analysis of economist Ronald Coase's 1937 article "The Nature of the Firm," you either paid the heavy costs of organizing or you didn't organize at all and got nothing done.

And so for centuries, people collaborated massively only on tasks that would make enough money to afford those costs. You could work together globally at building and selling profitable cars (like the Ford Motor Company) or running a world religion (like the Catholic Church), or even running a big nonprofit that could solicit mass donations (like UNICEF). Those organizations solved large, expensive, well-known problems—making cars, offering religion, helping the poor—that could generate serious cash flow. But what if a problem were smaller, more niche? Like how to find your way around a complicated video game? Well, there was basically no way to organize around it. Latent groups stayed latent.

Until the Internet came along. Now that self-organization online is basically free, those latent groups have burst into view. When a dozen friends spread across a city use a Facebook thread and a cute little voting app to pick which film they'll see on Friday night—"vote for your favorite!"—they are engaging in the same collective decision making that was previously available only to well-funded organizations. This, again, is basic behavioral economics: If you make it easier for people to do something, they'll do more of it. Finding your way around *Skyrim* or resolving conundrums like "Which movie are we seeing tonight?" are problems that traditionally couldn't afford Ronald Coase–style transactional costs—they fell "under the Coasean floor," as Shirky puts it. But things have decisively changed. "Because we can now reach beneath the Coasean floor," he writes, "we can have groups that operate with a birthday party's informality and a multinational's scope. . . . Now that group-forming has gone from hard to ridiculously easy, we are seeing an explosion of experiments with new groups and new kinds of groups."

The stuff that lives beneath the Coasean floor tends to be incredibly weird. I say that as a compliment. In a world where people think publicly and harness multiples to find like-minded souls, we can find other people worldwide who share our marginal interests, the sub rosa hobbies that we long nourished but couldn't find anyone geographically nearby to share. The result is a flood of amateur collaboration. I'm using "amateur" in its original sense, meaning not "done poorly" but "done for love instead of money." One of the reason some cultural elitists—political pundits, novelists, intellectuals—tend to be so unsettled by the Internet is that it has revealed how oceanically broad are the interests of the public in general. Before the Internet, with no way of observing the obsessions of the masses, it was a lot easier to pretend that these obsessions simply didn't exist; that the nation was "united" around caring about the same small number of movies, weekly magazines, novels, political issues, or personalities. This was probably always a self-flattering illusion for the folks who ran things. The Internet destroyed it. When you gaze with wild surmise upon the Pacific of strangeness online, you confront the astonishing diversity of human passion.

Take, for example, fan fiction. This is the art of writing stories based on one's favorite cultural products, such as novels, movies, or TV shows. It's an old practice, dating back at least to the early twentieth century, but in the 1970s, new variants began to emerge, spinning off from shows like *Star Trek*. One vibrant subset was slash fiction, written by *Star Trek* fans intrigued by the idea of a homoerotic relationship between Kirk and Spock—a wild, Heathcliffean hothead paired with a cool, distant partner. They started writing stories in which Kirk and Spock had an actual relationship, including plenty of steamy sex scenes, sometimes circulating the stories in photocopied zines. Pretty soon fans, very often women, were penning similarly riotous tales of other male couples from mainstream TV, including Starsky/Hutch pairings or Blake/Avon pairings, from the show *Blake's 7*. (Hence "slash" fiction, for the "/".) Other forms of fan fiction produced heterosexual couplings: *The X-Files* produced an avalanche of stories exploring the romantic life of Mulder and Scully, and Harry Potter produced Hermione/Harry tales. And plenty of fan fiction today has no sex at all; some fans just enjoy writing and love tinkering with their favorite fictional universes. The literary form gained some mainstream prominence in 2012 with the publication of *Fifty Shades of Grey*, a book whose author originally developed the story as *Twilight* fan fiction.

It's a subculture rife with multiples: finally all those folks in Iowa who'd been secretly penning tales based on *CHiPS* could discover those in Germany doing the same thing. These fans have been at the forefront of using digital tools to connect with one another. They became early expert users of bookmarking sites, putting tags on their stories—codes highlighting specific qualities—so readers could find exactly what they wanted. "If you wanted to read a 3000 word fic where Picard forces Gandalf into sexual bondage, and it seems unconsensual but secretly both want it, and it's R-explicit but not NC-17 explicit, all you had to do was search along the appropriate combination of tags," as Maciej Ceglowski, the founder of the bookmarking site Pinboard, wrote in his blog.

In 2011, Ceglowski had noticed a surge of fan fiction folks using his service. Since this was a profitable new base of customers, he figured he should find out if they had any requests. Should he add new features to Pinboard? On September 28 at 4:19 p.m. Pacific time, he tweeted an open-ended request: "Fanfic people, can you draft a list of your 'must have' features for me to look at, maybe as a Google doc? I'll implement what I can for you."

What resulted was an astonishing display of collaborative thinking. The fan fiction community instantly began retweeting Ceglowski's invitation. Within minutes, dozens from around the world had set up an open Google Docs document and began writing a wish list of features. That list quickly sprawled to thousands of words and became unreadably complicated, so the fans crowded into the "chat" area for the document and began editing it. One fan sorted the most common requests; another began carefully formatting the document to make it easier to read; others corrected typos. Within two days, they'd written a meticulously crafted sixteen-thousand-word "design document" for Ceglowski—mapping out how Pinboard could evolve, including samples of code he could use. He was astonished. "These people," he wrote later, "do not waste time."

The fans were almost as surprised as Ceglowski at how quickly and smartly they worked. "No one was in control. People were going, 'Where did all these people *come* from?'" as Priscilla Del Cima, a twenty-five-year-old graduate student from Rio de Janeiro who worked on the document for forty-eight hours, pausing only for brief snatches of sleep, tells me. There were so many people crowded into the chat area that it crashed ("Google Docs can't handle more than fifty people chatting at a time"). She's not sure precisely how many fans were involved overall—hundreds, she guesses—but the group was diverse enough that it tapped into a broad range of skills. "There were some very tech-savvy people on the chat, so when we started asking for something really hard they could say, 'Okay, *that's* technologically feasible and *that's* not.'"

This breadth of participation is key to what author James Suro-wiecki dubbed "the wisdom of crowds." Crowd wisdom as a scientific phenomenon was first explored in 1906 when the British scientist Francis Galton visited a county fair and observed a contest to guess the weight of an ox. About eight hundred fair attendees put in guesses. Galton expected that compared to the guess of an expert judge of oxen, the crowd of attendees would be far off the mark. But when Galton averaged out the crowd's guesses, the result was

1,197 pounds—just one pound less than the actual weight. "In other words," Surowiecki writes, "the crowd's judgment was essentially perfect."

How could that be? It's because, Surowiecki argues, each member of a decent-sized crowd of people possesses some incomplete part of the picture in her head. If you have a mechanism to assemble the various parts, you can wind up with a remarkably *complete* picture. The reason we never knew this was that for centuries, we had few mechanisms for assembling people's collective judgments. The ones that existed were large and professionally run, like polling firms or governments running elections. Now that ordinary people have mechanisms for aggregating knowledge, ever more complete pictures are swimming into view. The scads of Pinboard memo contributors each had one idea for improving the service; the thousands of *Skyrim* players each observed one small, stray fact about the game.

We see these mass think-ins happen frequently in response to cultural challenges, because people share so many cultural passions today. But group brainpower just as easily coalesces around difficult political problems, too.

Take the remarkable case of Tahrir Supplies in Egypt. In the fall of 2011, the Egyptian activists who had driven Hosni Mubarak from power faced a new danger: the Egyptian military. The military had assumed control and wasn't showing any signs of ceding to democratic control. So the protesters again massed in Cairo's Tahrir Square, where their movement had originally broken out. In November, on Mohamed Mahmoud Street just east of Tahrir Square, security forces moved in, unleashing torrents of tear gas and heavy beatings, even firing upon the protestors with live ammunition. The activists set up ten makeshift tent hospitals to treat the wounded.

Sitting in his apartment far away in Dubai, a twenty-three-year-old Egyptian named Ahmed Abulhassan was watching the fight via online video clips and Twitter. He had recently left Egypt after graduating from university with a degree in pharmacy biotechnology, but he knew dozens of friends who were back home being attacked by the military. "I was following the news and getting infuriated by it," he tells me.

But Abulhassan also noticed that the tent hospitals had a coordination problem. Doctors and volunteers with supplies would show up at one tent, but they'd actually be needed at a tent several blocks away. In the confusion, it was hard to know where to go. The activists did not have good "situational awareness," and Abulhassan realized he could use online tools to provide it.

He started a Twitter account called @tahrirsupplies, devoted solely to reporting which tent hospitals needed supplies. To get the word out, he begged various Egyptian celebrities and popular Twitter activists with hundreds of thousands of followers to retweet him.

It worked. Within a day, @tahrirsupplies had ten thousand followers, and activists were flooding Abulhassan with hundreds of texts and phone calls and tweets. He would learn which hospitals had which needs, then turn around and tweet that information. Doctors would know where to go, while Egyptian citizens would dash to their nearest pharmacy, buy supplies, and rush them over.

Abulhassan created, in essence, a highly collaborative, on-the-fly aid organization. It was like Médecins Sans Frontières, except without any head office or staff at all. As the battle raged for six days, requests got bigger. Three Egyptian women in their early twenties offered to help in the coordination. None were actually physically present; one lived in England, and two lived in Cairo but had parents who forbade them from joining the street action. "They said, 'Can we help?'" Abulhassan says, "and I said, 'Absolutely! This is not a business. I'm not *owning* it.'" Over the four days, the group slept in shifts, to make sure someone was awake to tweet requests.

Soon the requests were so numerous, they set up a Google Docs speadsheet to track the array of necessary goods, like Hemostop to seal wounds and Ventolin spray to heal lungs scarred by tear gas. The team even received requests for electrical generators and "an eye surgery machine for anesthetizing eyes." (Sure enough, wealthy Cairo residents bought and delivered those, too.) When pharmacists visited the hospitals to take an inventory, they estimated Tahrir Supplies had helped coordinate the delivery of one million U.S. dollars' worth of medical supplies.

Abulhassan never saw a penny, of course. Technically, his organization didn't exist. It was just a vehicle for helping people to help each other.

After the battle died down, Tahrir Supplies kept going, using its network to help out with other issues. It connected patients with rare blood needs—"we find people who are O negative and they both contact each other," Abulhassan says—and created medical caravans for poor villages. And when conflict broke out in Tahrir Square over the next year, it set about organizing aid there again.

In early 2012, Abulhassan found his first job after graduating. During the interviews, they asked him the standard interview question: How did you solve a problem in an innovative way? He laughed. For the first time, he had a good answer.

Collaboration and collective thinking don't happen by magic. In fact, they follow several rules, which you can see in action in these success stories.

First, collective thinking requires a focused problem to solve. With Tahrir Supplies, Abulhassan was trying to answer a clear question: Which tent hospitals in Cairo need help, and what do they need? This made it easier for a mass of people to contribute information. Indeed, he specifically avoided using his Twitter stream for anything *but* straightforward, verifiable factual information about the status of each hospital. (He didn't post thoughts about politics, even though, with a huge audience, he was tempted. "People were asking, 'What's your message?' But we were just trying to do one job.") Similarly, the contributors to the *Skyrim* wiki are united around a desire to produce a list of facts that will help players navigate a complex game.

Without a clear goal, collective thinkers can zoom wildly off base. Consider the spectacular collapse of the *Los Angeles Times*'s "wikitorial." Inspired by Wikipedia, the paper published a one-thousand-word editorial about the Iraq War as a wiki, then encouraged readers to edit it. "Do you see fatuous reasoning, a selective reading of the facts, a lack of poetry?" the *Times*'s editors wrote. "Well, what are you going to do about it? . . . Rewrite the editorial yourself." Within a day, the project began to unravel. Several hundred readers tweaked the text, but they couldn't agree on what direction to take it. Several made the tone more harsh: "The Bush administration should be publicly charged and tried for war crimes and crimes against humanity," one editor amended, and another simply wrote, "Fuck USA."

The problem with the wikitorial was that the goal had no obvious end point. A large group can argue about a set of facts and come to a reasonable consensus; Wikipedia does this every day. But a strongly worded *opinion*—the core of an op-ed—is not subject to consensus. This is why collective thinking online also tends to fail when it attempts an aesthetic creation. The Web designer Kevan Davis set up an online experiment to group-think pictures, allowing anyone to vote on the color of a randomly located individual pixel. But when the group tried to draw relatively simple figures—a castle, an apple, a human head, a cat—it produced undifferentiated blobs. Nobody could agree on what the subject ought to look like. Which is the point: Art is usually the product of a single independent vision. As most corporations discover to their dismay, groups can suck creativity out of projects because they tamp down the most original, idiosyncratic parts of each individual's vision.

Collective puzzle solving also requires a mix of contributors. Specifically, it needs to have really big central contributors—and then a ton of people making microcontributions.

It needs central people because, despite often being called leaderless, these projects are rarely entirely so. In reality, they almost always rely on a core group of contributors—the folks who frame the problem and get the project rolling. With Tahrir Supplies, it was Abulhassan; with the Pinboard project, while no one appointed themselves the leader, there were about a dozen very committed fans (Del Cima was one) who did crucial work like formatting the document and weeding out duplicate ideas. These hard jobs require enthusiasm and a time commitment, so they have a higher barrier to entry.

But they're not enough. Really successful collective thinking also requires dilettantes—people who offer a single small bit of help, like doing one edit or adding a fact or photo. Though each microcontribution is a small grain of sand, when you get thousands or millions you quickly build a beach. Microcontributions also diversify the knowledge pool. If anyone who's interested can briefly help out, almost everyone does, and soon the project is tapping into broad expertise: "The small contributions help the collaboration rapidly explore a much broader range of ideas than would otherwise be the case," as the author Michael Nielsen notes in *Reinventing Discovery*, an investigation of crowdsourced science.

Tahrir Supplies leveraged microcontributions brilliantly. Because Abulhassan made it so easy for activists to contribute news—all they needed to do was text, tweet at, or e-mail the central team— thousands did, and together they amassed a more complete picture of the situation than the central team could have achieved on its own. We see this pattern at Wikipedia, too. Because it's so easy to add to a Wikipedia article—hit "edit" and boom, you're a contributor—the scale of microcontributions is vast, well into the hundreds of millions. Indeed, the single most common edit on Wikipedia is someone changing a word or phrase: a teensy contribution, truly a grain of sand. Yet, like Tahrir Supplies, Wikipedia also relies on a small core of heavily involved contributors. Indeed, if you look at the number of really active contributors, the ones who make more than a hundred edits a month, there are not quite thirty-five hundred. If you drill down to the *really* committed folks—the administrators who deal with vandalism, among other things—there are only six or seven hundred active ones. Wikipedia contributions form a classic long-tail distribution, with a small passionate bunch at one end, followed by a line of grain-of-sand contributors that fades off over the horizon.

These hard-core and lightweight contributors form a symbiotic whole. Without the microcontributors, Wikipedia wouldn't grow as quickly, and it would have a much more narrow knowledge base. (And Wikipedia's base of microcontributors still needs to become even more diverse, frankly. Because Wikipedia's hard-core contributors tend to be mostly techie white men; it has terrific coverage of computer science and physics—but big holes in other areas. "I go to find an article about Swedish feminism because I'm reading Stieg Larsson, and it doesn't exist," as former Wikimedia Foundation executive director Sue Gardner once told me; she's been trying to recruit more women and non-Western contributors.)

This blend of microcontributors and heavy contributors works well for gathering and organizing data. But it even works with "insight" problems—the type that require an aha breakthrough. Consider again the world of chess, where such epiphanies are critical. In 1999, Garry Kasparov engaged in yet another fascinating experiment by playing a game against an online collective. Billed as

"Kasparov versus the World," it allowed anyone interested in the game to visit a Web site where they could suggest a move and vote on the best one to play against Kasparov. More than fifty thousand people participated from more than seventy-five countries, with an average of five thousand people voting on each move. The group contained a few heavy contributors, most notably the fifteen-year-old Irina Krush. A rising star in the chess world, Krush had recently become the U.S. women's chess champion. At move 10, Krush suggested a play—which the collective adopted—so powerful that Kasparov called it "an important contribution to chess." As Nielsen points out in *Reinventing Discovery*, Krush also helped coordinate the mob: Together with her management team, she created an online "analysis tree" that listed possible gambits, so the group could stay on the same page as they voted.

But the microcontributions and cognitive breadth were crucial. Having one really smart player, or even a small handful, wasn't enough to fight Kasparov. Krush said one of her three favorite moves the team played was number 26, which came not from her but from Yaaqov Vaingorten, "a reasonably serious but not elite junior player." Thousands of other players offered tiny bits of analysis that shaped successful plays. In the end, Kasparov still triumphed, but it took him sixty-two moves and several nail-biting reversals of fortune. Collectively, the group played far more skillfully than Krush or any of the group's individual players could. Kasparov called it "the greatest game in the history of chess," adding, "The sheer number of ideas, the complexity, and the contribution it has made to chess make it the most important game ever played."

There's one final, and very subtle, part of smart collective thinking: culture. It turns out that the type of people in the group and the way they interact spell the difference between success and failure.

Wikipedia is perhaps the most famous collaborative project. The real secret to its success, though, isn't merely its millions of volunteers. It is, as communications professor Joseph Reagle dubs it, the culture of "good faith collaboration" that Wikipedia cofounder Jimmy Wales labored to put in place—a commitment to Quaker-level civility. Not long after launching Wikipedia, Wales penned an open letter to all potential contributors (by which he meant the entire planet), arguing that the project would only work if the contributors struggled constantly to remain polite to one another. "Mutual respect and a reasonable approach to disagreement are essential . . . on this incredible ridiculous crazy fun project to change the world," Wales wrote. This attitude was later codified by the users themselves as one of Wikipedia's Five Pillars of self-governance:

> Respect your fellow Wikipedians, even when you disagree. Apply Wikipedia etiquette, and avoid personal attacks. Seek consensus, avoid edit wars, and never disrupt Wikipedia to illustrate a point. Act in good faith, and assume good faith on the part of others. Be open and welcoming.

Another of Wikipedia's Five Pillars addresses the site's "neutral point of view": "We avoid advocacy and we characterize information and issues rather than debate them." Indeed, faced with a controversial subject about which she feels strongly, a Wikipedia contributor ought to work extra hard to carefully describe views she finds repellent. Since Wikipedia contributors regularly disagree about facts—ranging from hot-button issues like "when life begins" on the abortion page to whether *Star Wars* media ever actually identify Yoda's home planet—the only thing keeping articles from being endlessly rewritten by warring factions is for the factions to stop warring. That can't be done by software; it takes culture. As Reagle points out, Wales spent countless hours illustrating his own principles by politely urging combatants to be, well, polite. To defuse the sense of urgency that often makes arguments more bitter, Wales would point out that actually there's no rush at all in working on Wikipedia, and in fact "there is plenty of time to stop and ask questions." The upshot, Reagle notes, is that the interactions among Wikipedians often continue at prodigious length. They go on seemingly forever, which is a deficit (they can be mind numbing) and a delight (they permit and encourage the sort of crazy rathole exploration that leads to productive thinking). They are "frequently exasperating, often humorous, and occasionally profound," as Reagle writes. Even white supremacists who've tried to edit Wikipedia pages—surely among the most adversarial, antagonistic contributors one could imagine—have absorbed this culture and "reminded themselves they need to be cordial on Wikipedia," as Reagle has found.

To be really smart, though, an online group needs to obey one final rule—and a rather counterintuitive one. The members can't have *too much contact* with one another. To work best, the members of a collective group ought to be able to think and work independently.

This rule came to light in 1958, when social scientists tested different techniques of brainstorming. They posed a thought-provoking question: If humans had an extra thumb on each hand, what benefits and problems would emerge? Then they had two different types of groups brainstorm answers. In one group, the members worked face-to-face; in the other group, the members each worked independently, then pooled their answers at the end. You might expect the people working face-to-face to be more productive, but that wasn't the case. The team with independently working members produced almost twice as many ideas. Other studies confirmed these results. Traditional brainstorming simply doesn't work as well as thinking alone, then pooling results.

That's because, the scientists found, groups that have direct contact suffer from two problems. The big one is blocking—a great idea pops into your head, but by the time the group calls on you, you've forgotten it. The other is social dampening: outspoken, extroverted members wind up dominating, and their ideas get adopted by others, even if they're not very good ones. Introverted members don't speak up. In contrast, when group members work physically separately from one another—in what researchers call "virtual groups"—it avoids this problem because everyone can generate ideas without being cognitively overshadowed or blocked. This is one of the counterintuitive secrets behind online collaborations. They inherently fit the model of people working together intimately but remotely.

If we're too readily swayed by the views of people when we're in a room with them, is it possible to be similarly led astray by the views of people online? Apparently so. A 2011 study took several virtual groups and asked them to estimate obscure facts about Switzerland, such as its population density and crime rate. At first, the group members had no exposure to each other, and—like the folks guessing the weight of the ox—the group was smart. The average of their guesses was close to the facts.

Then the scientists changed the experiment. As the group members were cogitating alone in their cubicles, the scientists gave them more information: Each member could see the others' guesses and the average of the group and then they could revise their guess. What happened? The wisdom of the group broke down. Members began to drift toward each other, influenced by the others' guesses. This caused a bad feedback loop, because as each individual member revised his own guesses, the overall average guess grew less accurate. By being exposed to one another's individual thinking *too* intimately, the group's collective wisdom declined. It got stupider. The members began to suffer from precisely the same problem that can afflict face-to-face groups.

This problem, of course, crops up all the time online—the "rich get richer" cycle of popularity. When a newspaper puts a list of its most-e-mailed stories on its front page, it represents the collective judgment of its readers. But it's also biasing incoming readers. The crowd knows too much about what the crowd thinks, and its wisdom vanishes in an Ouroboran gulp of its own tail.

Now, obviously we're not completely robotic lemmings, otherwise the top-ten lists would *never* change. Even when exposed to one another's views, we still hold our own counsel. But this question— "How much should an online crowd know about itself?"—turns out to be among the biggest design challenges for anyone trying to harness collective thinking. The trick is to encourage people to join in but also to think for themselves.

Historically, one way we measure intelligence is by evaluating one's ability to solve problems. But in collective thinking, a new proposition emerges that flips this logic on its head. The key skill here is *designing problems*—designing them in a way that lets many people pitch in to solve it.

This is the way David Baker and his team cracked a ten-year-old biological puzzle. Baker is a biochemistry professor at the University of Washington who studies proteins, the devilishly complicated molecules that make up the human body. A protein is an incredibly long string of atoms folded into a tight ball—think of a long piece of yarn rolled tightly into a ball around itself. There are billions of possible ways to fold up a protein into a ball, each making the protein behave in a slightly different fashion. If you can figure out how the protein folds, you can understand how it works, and this is crucial in both understanding diseases and designing drugs to combat them.

For years, Baker had been studying protein folding with brute-force computer power. He'd create a computer model of a long protein string, then have his computer semirandomly fold it, hoping to hit upon a revealing result. The process was glacially slow; searching randomly is not very efficient. To amass more computing power, Baker created Rosetta@Home, a distributed computing application. You could download the program and run it as a screen saver; if your computer was sitting idle, it would start randomly folding proteins, effectively contributing free computer power to Baker's academic research. Thousands of people downloaded Rosetta@Home, eager to help out the cause of science. It was also quite beautiful to watch, because you could see a graphic of the protein being folded on-screen.

Then Baker started getting e-mails from users. They'd watch the protein being folded and realize, Hey, I could do this better than the computer. The computer was dumbly trying random combinations. Humans could spy better ones. They wanted to try doing it themselves.

A collaborative-thinking project was born. Baker joined forces with some programmers at his university and created Fold.it, a program that let people fold a protein themselves. Users employed the mouse and buttons to tweak, jiggle, and twist proteins into increasingly elegant balls. Fold.it had a fun, gamelike interface: If you managed to produce an efficient fold, it would generate a high score. Pretty soon there were two hundred thousand players collaborating. They set up a wiki to discuss their favorite strategies, cramming into the Fold.it chat rooms to share tips. Some formed teams, with names like Void Crushers or Contenders, to achieve high scores. "Encouraging discussion and questions, all are free to express themselves," as the Contenders explain on their Web site. "We play our soloist games our own way; but if someone finds sudden success, it's posted for the benefit of the group, detailing what was done to get there." In effect, Baker had designed a problem that perfectly leveraged collective thinking. It had a specific goal, brought in a diverse range of expertise, encouraged microcontributions, and let people think quietly on their own even while collaborating with others.

Soon, it led to scientific breakthroughs. In 2010, Baker decided to give the Fold.it community a particularly tricky challenge: folding the M-PMV virus, which causes AIDS in monkeys. Biologists had been trying for a decade to figure out how M-PMV folds but had arrived at only an incomplete solution. The Fold.it players quickly began improving on that solution, with one finding a cleverer fold, which another one would improve upon, and so on.

In only three weeks, the amateurs of the Void Crushers and Contenders teams had jointly solved a problem that had bedeviled professional biologists for ten years. Baker published the results in *Nature Structural & Molecular Biology* in an essay titled "Crystal structure of a monomeric retroviral protease solved by protein folding game players."

Baker is continuing to use the Fold.it community to solve hard folding problems; in fact, he has published three more papers based on their successes. In one case, they came up with a "crazy" way to fold a binding enzyme that Baker and his colleagues had never considered. In another paper, Baker studied the strategies players were using to fold proteins and found they had independently hit upon techniques that biologists had been using in private but hadn't published yet. How had the crowd been so innovative? Again, by being very open and sharing their best work and posting their favorite techniques so others could improve them. "The best ones would just go viral," Baker tells me. Baker's success wasn't in finding a solution. It was in designing a good problem—a game system that neatly channels the capabilities of the group.

Motivation matters. People eagerly pitch in on projects they're interested in, which is exactly why so many ad hoc collaborations erupt around amateur passions, like hobbies or pop culture. But as the Fold.it guys found, science projects work well, too, because people enjoy feeling they're contributing to global knowledge. Many other scientists have joined Baker in crafting successful group thinking projects, such as the Galaxy Zoo, founded by the Radcliffe Observatory in Oxford, England, which puts a deluge of space imagery online and lets everyday astrophiles classify the shapes of galaxies; like Fold.it, it quickly evolved a community of contributors. Politics, too, is an area where many people are motivated to help, which is what drives the success of the Ushahidi maps or government 2.0 projects, where citizens compile information to improve their communities.

But can you make *money* off collective smarts? Can they help corporations work more intelligently?

It's harder than you'd think. Compared to the many public-minded projects, few corporations have been able to harness huge, public groups of collective thinkers. Motivation is a problem: Few people think profitable companies deserve their free work. At best, companies have been able to deploy fairly simple polling-and-voting group thinking projects, often to tap in to what their customers want; clothing firms like Threadless let their users vote on user-submitted designs. Others have solved the motivational problem by offering substantial prizes. Netflix, for example, offered a one-million-dollar prize for whoever could improve its movie-recommendation algorithm by 10 percent. But while prizes motivate hard work, they can inhibit sharing. When people are competing for a big prize, they're often not willing to talk about their smartest breakthrough ideas for fear that a rival will steal their work. (Indeed, as teams got closer to winning the Netflix prize, they became increasingly secretive.)

Other corporations have solved the problems of motivation and secrecy by turning inward and creating internal "decision markets" where employees can pose ideas and vote on the best ones. At one

software firm, Rite-Solutions, an employee pitched a new product—
a 3-D environment designed to help military clients handle
emergencies—and after it got heavily upvoted, the firm decided to
build it, whereupon it became the source of 30 percent of their annual sales. Internal markets can be extremely valuable, because they
keep trade secrets secret. But they dramatically shrink the pool of
people who help solve problems. As Tahrir Supplies discovered,
when it comes to microcontributions, scale matters. If, as Sun Mi-
crosystems cofounder Bill Joy reportedly liked to say, the smartest
people are outside the room, collective problem solving requires being as public as possible. That's harder with for-profit challenges.

The most successful corporate think-ins emerge when consumers
perceive an overlap between their self-interest and the self-interest
of the corporation. Some newspapers have tapped crowd smarts to
assist with investigative journalism, because enough people feel
these investigations are a public good. If designed well, they're also
fun. When the UK government was forced to release a trove of expense receipts for British politicians, the unsorted pile was too big
for any single journalist to sift through—so the *Guardian* created a
gamelike online tool that let anyone comb through the stack and
flag dodgy-looking expenses. More than twenty thousand citizens
analyzed a stunning 170,000 documents in four days, and the
Guardian published a story listing some of the crowd's most egregious discoveries (such as an MP who'd charged £225—$441—for
a duvet.) Several political sites have used their large passionate audiences to comb through similar governmental data dumps. And to
figure out which neighborhoods charge the most for groceries, fans
of the Brian Lehrer radio show in New York sent in the price of let-
tuce, beer, and milk in their local markets, which produced a witty
map of the city.

Google, which relies on analyzing links, has essentially built its
search empire on collective knowledge. Every time someone posts

a link to a Web site, they're giving Google information to analyze; each link is a tiny vote for the site's relevance. Several other Google projects have leveraged different types of collective effort. When I visited the offices of Google Earth, its product manager, Peter Birch, booted up the software and zoomed in to Red Square in Moscow. As we approached street level, I could see hundreds of buildings appear, perfectly modeled in 3-D, including gorgeously rendered versions of St. Basil's Cathedral, with its colorful, bulb-topped towers. Google didn't design those buildings; fans of 3-D modeling did. Google simply made it easy to contribute, releasing free Building Maker software and an online tool for submitting your building for inclusion in Google Earth. If it's accepted, Google includes your user name in the model, so people can know who made it and see all your other buildings, too. "Now we have an amazing amount of buildings all over the world," Birch told me, hovering his mouse over different buildings to show who'd crafted them. "And who knows where these people are and where they live? But that's the kind of cool thing about it. People are able to com-municate through this tool, where they can share all this informa-tion." Google Earth's relative openness—and its value as a creative showcase for one's 3-D-modeling skills—turned out to be a tempt-ing invitation to contribute, even though Google is clearly a for-profit entity. (In 2013, Google launched an even faster way of generating buildings—by using satellite photo data—and retired the building-maker tool, though it kept many of the buildings created by contributors.)

Still, because openness is most natural in amateur work, I sus-pect the leading edge of collective thinking—as with Wikipedia or Linux—will always emerge in the amateur world. If you want to see the future of collective thinking, don't watch what Fortune 500 firms are doing. Watch what fan fiction writers are doing or what activists are doing. Or even watch how smart individuals do it—the ones who cultivate broad, diverse networks of friends or followers online.

The potential of collective thinking is only going to grow. Each new tool for expression, each new vehicle for talking to one another, opens up new potential forums for collaboration. Some regard this as alarming: Critics have complained that the online "hive mind" is dehumanizing. And it's true that thinking of people as bees in a hive, devoid of agency, is rather depressing. But this is precisely why the hive metaphor isn't all that accurate. Humans are not ants, as the philosopher Pierre Lévy noted in his 1994 book *Collective Intelligence*. We participate in larger groupings when there's something there that enhances our individual humanity. The collaborative thought projects that succeed are the ones where each act of participation, no matter how small, excites and rehumanizes us. This is why Wikipedia succeeds. Those who make an edit or participate in a talk-page discussion transform their sense of self, becoming creators of knowledge.

Personally, I think there's a better metaphor for collaborative thinking: Sherlock Holmes. Arthur Conan Doyle's detective is famously brilliant. But he's also famously bored. When he doesn't have a complex case, something so fiendishly difficult that it engages all of his electric intellect, the boredom drives him nuts. He shoots cocaine ("a seven-per-cent solution") and begs for the universe to send him a case complicated enough to engage his analytic abilities.

"My mind rebels at stagnation," he tells Watson in *The Sign of the Four*. "Give me problems, give me work, give me the most abstruse cryptogram, or the most intricate analysis, and I am in my own proper atmosphere. I can dispense then with artificial stimulants. But I abhor the dull routine of existence. I crave for mental exaltation." Or in *The Adventure of Wisteria Lodge*: "My mind is

like a racing engine, tearing itself to pieces because it is not connected up with the work for which it was built."

This is what the enormous, latent collaborative intelligence of
humanity is like: one big Sherlock Holmes, craving problems that
suit its peculiar powers. Like the networked video-game players, it's
hungry for harder puzzles. We have to learn how to design them.

Digital School_

When I visit Matthew Carpenter's math class, I peer over his shoulder at his laptop and see on-screen the question he's tackling:

$$cos^{-1}(1) = ?$$

It's a scrap of inverse trigonometry. I've long forgotten much of my trig, so I shrug. Matthew, however, is undaunted. Squinting in concentration, he clicks one of the four possible multiple-choice answers: 0 degrees. *Ding:* the software informs him that he's gotten it right. It throws another question at him, which he also answers correctly; then another, and another, until he's aced ten questions in a row. "This is my favorite exercise right now," he tells me. He's certainly practiced a lot. He points to a section of the screen that shows he's tackled 642 inverse trig problems. "It took a while for me to get it," he admits sheepishly, but he's plugged away at it in class and at home for hours.

Matthew shouldn't be doing work remotely this advanced. He's ten years old, and this is only the fifth grade. Matthew is a student at Santa Rita Elementary, a public school in Los Altos, California, where his sun-drenched classroom is festooned with a giant paper X-wing fighter, student paintings of trees, and racks of kids' books. Normally grade five math is simpler fare—basic fractions, decimals, and percentages. You don't reach inverse trig until high school.

But Matthew's class isn't typical. For the last year, they've been using the Khan Academy, a free online site filled with thousands of instructional videos that cover subjects in math, science, and economics. The videos are lo-fi, even crude: about five to fifteen minutes long, they consist of a voice-over by Khan describing a mathematical concept or explaining how to solve a problem while hand-scribbled formulas appear on-screen. The site also includes software that generates practice problems, then rewards hard work with badges—for answering a "streak" of questions right, say.

Matthew has amassed fifty-two Earth badges, one of the more desirable awards on the site. When a girl in a pink jumpsuit wanders by and peeks at Matthew's recent streak of inverse trig, she groans: "Oh, great. I need to catch up with you now!" She flops down at a laptop and begins pecking away at her own trigonometry problems. Her first one asks her to divine the slope of $y = -1x^3 + 4y^2$; scribbling on a piece of paper as I watch, she figures it out in a few seconds and clicks the right answer on screen. So far, she tells me, she's watched dozens of hours of math videos, mostly at home.

How did these elementary school kids zoom ahead to high-school-level material?

In part because the site lets them learn at their own pace—allowing their teacher, Kami Thordarson, to offer much more customized instruction. The problem with traditional classroom dynamics, Thordarson tells me, is that they don't easily account for the way kids learn at different rates. When she stands up at the chalkboard lecturing on a subject, there's a predictable pattern that takes hold: one quarter of the kids quickly fall behind, so they tune out. Another quarter already know the material, so *they* tune out. At best, "you're teaching to this middle group of students." Thordarson sighs.

What works better? Personalized, one-on-one tutoring. Back in 1984, the educational scholar Benjamin Bloom compared students taught in regular classrooms—one teacher lecturing to the assembled class—to students who got months of one-on-one attention or instruction in small groups. These tutored students did far better; two standard deviations better, in fact. To get a sense of how much of an improvement that is, think of it this way: If you took a regular-classroom kid who was performing in the middle of the pack and gave her one-on-one instruction for a few months, she'd leap to the ninety-eighth percentile. This became known as the "Two Sigma" phenomenon, and in the decades since, public-school teachers have struggled to give students more one-on-one time. This isn't easy, given that the average class in the United States has roughly twenty-five children. (Worse, after years of slightly falling, that number is now rising again, due to budget cuts.) Until the government decides it's willing to subsidize smaller classes, how can teachers get more personal time?

One way is by using new-media tools to invert the logic of instruction. Instead of delivering all her math lessons to the entire class, Thordarson has them watch Khan videos and work on the online problems. This way, the students who quickly "get it" can blast ahead—and Thordarson can focus more of her class time on helping the students who need coaching. Other teachers are even more aggressive about inverting their classes: They assign videos to be watched at home, then have the students do the homework in class, flipping their instruction inside out.

This makes curious psychological sense. A video can often be a better way to deliver a lecture-style lesson, because students can pause and rewind when they get confused—impossible with a live classroom lesson. In contrast, homework is better done in a classroom, because that's when you're likely to need to ask the teacher for extra help. (Or to ask another student: Thordarson and her colleagues noticed students helping one another, sharing what they'd learned, and tutoring each other.)

"Kids get to work in their place where they're most comfortable," says Thordarson as we wander around her class. "They're allowed to jump ahead. It gives kids who are above grade level a chance to just soar! And for kids who struggle, it gives them a chance to work through some of those issues without everybody watching."

Still, as Thordarson quickly points out, the Khan Academy isn't enough on its own. You can't just plunk kids in front of laptops and say, "Go." The point isn't to replace teachers. It's to help them reshape their classes in new ways—and spend more time directly guiding students. You can't even say it makes the teacher's job easier. If anything, it has made Thordarson's job *more* challenging: there's more noise, more kids talking, and she's constantly darting around the room to help out. One U.S. federal study found that students learned best in classrooms with precisely this sort of "blended" learning—traditional teachers augmented with online instruction. But the increase in learning wasn't because of any magic in the medium. It's just that online tools helped students and teachers spend more time on the material.

Judging by Thordarson's success, though, it works. She's seen particularly strong improvements at the low end: Only three percent of her students were classified as average or lower in end-of-year tests, down from thirteen percent at midyear—and other math teachers at Santa Rita have seen similar results. The kids who need help have been getting more of it; the kids who want to push ahead are pushing ridiculously far.

"It's like having thirty math tutors in my room," Thordarson says.

The classroom hasn't changed much over the years. Over the centuries, actually. In the 1350s, artist Laurentius de Voltolina painted a scene of a university lecture in Bologna that looks quite like a present-day classroom: The professor sits at a podium at the front,

pontificating to twenty-four seated students, one of whom is keel-
ing over in apparent boredom, four of whom are ignoring the lecture while talking, and one of whom appears to be completely
asleep. As various educational analysts have joked, if you brought a
bunch of surgeons from a hundred years ago into today's hospitals,
they would have no idea what was going on, because everything
about their craft had evolved: antibiotics, laparoscopic devices,
MRIs. But time-traveling teachers would have no trouble walking
into an elementary school (or even Harvard) and going to work,
because schools are nearly identical. Walk to the front of the class,
pick up the chalk, and start lecturing.

These days, a huge debate rages in the United States about how
schools ought to modernize. The school-reform movement argues that schools are "failing" because they're hobbled by union-
coddled teachers who block change. Only by rigorously testing kids
and firing teachers who can't produce rising scores—while offering
merit pay to teachers who can—will schools improve. Critics of the
reform movement counter that these standardized tests not only fail
to measure actual learning but deform education, as teachers drill
children in meaningless test-prep skills. To really improve schools,
goes the counterargument, you'd need to seriously train teachers (as
they do in Finland), tackle the poverty that haunts many students'
home lives, and pony up public funds for smaller class sizes.

This debate extends cosmically beyond this book, so I'm certainly not going to resolve it here. (Though I personally agree more
with critics of the school-reform movement.) I'm also not going to
delve deeply into the question of how frequently young students
ought to be sitting in front of screens, another fraught area of debate. There's little clear data on the subject as yet, though the work
I've seen suggests that children in early grades learn best with
hands-on, tactile experimentation—which suggests digital tools
ought to be used very sparingly early on. What interests me here is

the unique roles that digital tools might have in the later grades: *Can* they help students learn, and if so, how?

Of course, classroom technology has a long history of hype that has rarely delivered. In the nineteenth century, George Parsons Lathrop predicted that movies would let children experience "the majestic tumult of Niagara" or "the animated presence of far-off peoples." Soon after, radio became the hot new thing; surely it would enrich students with high-quality lectures. A couple of decades later, the journal *Nation's Business* declared that TV would become "the biggest classroom the world has ever seen." Each technology barely made a dent.

Of these newfangled devices, computers have been the most successful in infiltrating classrooms—certainly in terms of dollars spent. Schools have spent an estimated sixty billion dollars in the last twenty years on computers. Indeed, manufacturers and software firms love to target schools, since, rather like the military, they are lucrative customers, an enormous sector loaded with tax dollars to spend. (Apple computers are omnipresent in schools not just because they're well made, but because, as *New York Times* reporter Matt Richtel has documented, Apple salespeople actively woo school officials to visit their Cupertino headquarters, paying for hotel stays and pricey dinners.) As a result of this high-tech spending, the ratio of computers to kids has shifted from 1:12 in 1998 to less than 1:4 today.

Despite all that spending, computers are not often used to do anything that couldn't be done as easily—and more cheaply and effectively—with old-fashioned books, pencils, paper, and chalk. "Interactive" whiteboards are too frequently used merely for displaying text or pictures to a dutifully seated class, not much different from what you can do with an overhead projector or chalkboard. Computers are used for mere word processing, to display cognitively bleak commercial software packages like PowerPoint, or to play dull educational games of dubious merit. Computer "art" packages get used as bland substitutes for paint-and-paper art.

But some educators are realizing that this is a dead end. They're the ones like Thordarson, who are doing something different with technology in the classroom. These teachers have realized that the point isn't to simply replicate more expensively what they're already doing quite well with paper, pencils, and books. It's to do new things that they currently *can't* do. It's to teach kids by using the peculiar abilities of networked devices—like public thinking, new literacies, and the powerful insights that come from not just using, but programming, the machine.

Consider what's happening beneath the hood of the Khan Academy. In one sense, Khan's videos are the most prominent part of the system. But they're also the least innovative one. They're still pretty much just traditional lessons and lectures, albeit ones that can be consulted and reconsulted worldwide, at any time.

What's new is how teachers use the Khan Academy to track progress. The system offers a dashboard that displays nuanced information about each student: which videos they've looked at, which problems they've tackled, how many times they had to work at a problem before they solved it. This data offers pragmatic insight into whether a student is struggling or not—in real time, whether the child is working in the classroom or at home. It takes the invisible and sometimes mysterious development inside a student's head and makes it visible.

One afternoon I drop by the seventh-grade classroom of Courtney Cadwell, a math teacher at Egan Junior High in Los Altos, California, across town from Thordarson's elementary school. Cadwell is a high-energy former Texan who was in line for NASA's astronaut training program before becoming a teacher (and who has since become the principal of another Los Altos school). She needs that energy because her job is about as hard as they come: Her students are remedial, and many come from poor immigrant families in which the parents speak little or no English, so they're rarely able to help their children with homework. When the children arrived at the beginning of this year, some were functioning at a third grade level in math, she tells me.

The dashboard helped her zero in on what kind of help each student needed, often in real time, letting her tailor her instruction or offer extra guidance. "Usually we get to the end of a lesson and we get a quiz or a test and then you realize, Oh my gosh, they have all these gaps!" Cadwell says. "With this, I can spot it sooner and fill those gaps."

Being able to target students has paid off dramatically. In her first six months using the Khan Academy, her class's scores had improved more than 106 percent. One girl advanced an astonishing 366 percent. Whereas the class average used to be three years behind, now it's a bit over a year behind. At this rate, by the end of the year, Cadwell will have them caught up. She pulls up the dashboard to show me the charts. Interestingly, their progress goes in bursts: They seem to grapple with a concept for days or weeks, then suddenly get it and improve abruptly in performance.

Cadwell has been teaching for eighteen years but she's never been able to so quickly identify and address students' areas of weakness. "It's just incredible, these gains, you know!" she gushes. "It's *insane*."

I hang out and observe the students on their laptops as they watch videos and blast through problems while Cadwell darts around, coaching one and then another. I talk to the student who'd advanced by 366 percent, a smiling, long-haired girl in purple-rimmed glasses. "I hated math," the girl tells me cheerfully as she plows through a long-division problem. "But now it's actually fun." Like many students, she also found it encouraging to see her own improvement—to watch the chart line move upward. Normally students have only a crude sense of the rate at which they're learning, but when software is charting their progress, it's motivating simply to see the data. "When you're collecting badges for getting a streak of questions right, it's instant feedback," one of them tells me. "You want to keep going, doing more, more, more." One Santa Rita teacher told me she'd noticed that a kid had worked on problems from midnight until 2 a.m. during a bout of insomnia.

Collecting such fine-grained data is likely to have other payoffs. When I talk to Sal Khan, who runs the Khan Academy as a nonprofit supported by donations, he points out that students have answered more than a billion questions on his system, and the videos have been viewed over 230 million times. "So we can start looking for trend lines that help us figure out, What types of things are students likely to get stuck on? If someone breezes through trigonometry but gets stuck on the intro to statistics, can we predict what other things they'll find hard or easy? Can we help give more information to teachers to help them teach?" This is, of course, one of those things that computers are uniquely good at: finding patterns that we can't see ourselves.

The Khan Academy can work for math and sciences, where problem sets can be autogenerated and automatically graded. But what about teaching kids to read and write?

These skills lags behind math, believe it or not. One of the most reliable measures of how kids are learning in the United States is the National Assessment of Educational Progress, a testing program run by the federal government. The tests are reliable mostly because they're not used to assess or rank individual schools and teachers; since nobody's worried about getting fired if their kids don't excel on NAEP tests, the results aren't deformed by test prep. And what do they show? News both good and bad. The good part is the math scores, which have slowly and slightly risen over the last thirty-five years. The bad news: the reading and writing scores. Only the lower grades have gained; amazingly, the average reading ability of seventeen-year-olds in 2008 was nearly identical to what it was in 1971. As for writing, the NAEP's data on this isn't complete, since it was only collected for a decade beginning in the mid-eighties. But in that window, things didn't look great: Writing skills remained generally flat and got worse for kids in eleventh grade.

Studies show that the main way to help kids read is to give them more reading time in school, offer them books they're interested in, and teach them the mental strategies of good readers—like summarizing what they're understanding and what they're not.

But research also shows something else: *Writing* is curiously pivotal to reading. As I noted when I first discussed public thinking, writing about the stuff we're reading activates the generation effect. We internalize our reading more deeply. Indeed, literacy scholar Steve Graham recently crunched dozens of reading studies and found that "writing about a text proved to be better than just reading it, reading and rereading it, reading and studying it, reading and discussing it, and receiving reading instruction."

But as any teacher knows, getting students to write is unbelievably difficult, in part because writing assignments feel so artificial. After all, only the teacher is going to read your paper. Why bother working hard on it?

Dorothy Burt, a literacy project facilitator in Point England, New Zealand, knew all about these problems. Like Cadwell's school in California, Burt's is located in a low-income cluster of schools, with high illiteracy rates and many students for whom English is a second language. Back in 2007, the writing skills of students in that cluster languished far below the national average. The teachers wanted their kids to practice writing more but could only get a few sentences out of them.

They decided that the students were right. Composing essays *is* meaningless. Teachers are an inauthentic audience. They aren't necessarily interested in what their students have to say; they're just reading as a part of their jobs. A writer is being forced to write for an audience that's being forced to read. No wonder they think it's pointless, Burt thought.

So the New Zealand teachers decided to try using the power of public thinking. Instead of having assignments filed in paper and dropped on the teachers' desks, students would post on public blogs. Anyone could read the posts and comment on them, and the school encouraged parents and friends to do so, including those overseas.

At first, nothing changed. The students still grumbled. But soon, when comments started appearing, they realized that they were writing for a real audience. Occasionally the comment would come from a complete stranger—and *that* caught the students' attention. "How is someone from *Germany* reading what I'm writing?" one student said. When another student posted a review of a book, the book's author heard about it and showed up to comment, too.

The students were electrified. They began writing posts far more frequently, waiting patiently for computer time and groaning when teachers told them to stop. The ones who had computers at home posted in their own time, on weekends and vacation. Previously reluctant readers began doing more careful research. Students began paying closer attention to grammar and punctuation: They'd read their entries aloud to notice that commas were missing or used too often.

Students also began critiquing each other's writing, demanding they clarify their points. "I don't get what you mean," one told another. "We can't read your mind." Some began editing one another's posts to remove local jargon, reasoning that foreigners wouldn't understand the references to, for instance, New Zealand's national rugby team: "People in America won't know what the 'All Blacks' are," one student admonished another. "They were writing for a global audience," one local literacy researcher, Colleen Gleeson, tells me. These are acts of self-awareness that professional writers struggle with: forming a theory of mind of one's audience, the better to communicate with it. Or as Burt adds when I interviewed her: "The blogging environment gave the students an audience that had a choice not to read. So if they do choose to read it it's because they want to."

A year into the program, the New Zealand schools decided to expand the experiment dramatically, by finding a way to get every student a netbook. It cost each family about twelve U.S. dollars per month per student—significant, but affordable—and after three years the student would permanently own the laptop. The schools even began building neighborhood-wide WiFi networks to share their connection freely.

Did the experiment actually improve learning? When they looked around, Burt and the teachers could see dozens of positive effects. Students were more excited to write and research, and saw school as newly relevant—they were, in the au courant term, engaged. But "engagement" is a fuzzy, catch-all rationale for using technology. Sure, having lots of shiny new devices around might confer a sense of novelty, improving "engagement"—but parents and politicians aren't going to care unless you can actually prove the kids are learning. They want test scores to go up. (And while test scores are a crude way to measure actual education, they're the current yardstick, for better and for worse.)

It turns out the scores did improve, and dramatically. By the end of the second year, the scores of those New Zealand schools for reading and writing were advancing at an astonishing clip—some posting improvements ten to thirteen times larger than the national average increase. In 2008, they were dramatically below the nation at large. By 2011, some grades had reached parity, and some were above.

The motivational force of public thinking and online collaboration for kids is even easier to see if you look outside the classroom—because that's where it's mostly happening. As fan culture scholar Henry Jenkins once pointed out to me, children and young adults who bridle at writing a few paragraphs in school will cheerfully spend months on sprawling works of online prose, from fan fiction to TV show recaps to video game guides. One student I spoke to, Eric Davey, wrote a patiently detailed thirty-eight-page guide to a *Star Trek* game that clocked in at thirteen thousand words—while enrolled in a high school where the longest assignment was two to four pages. Over at the Lostpedia, one of the pioneers was a thirteen-year-old boy named Sam McPherson, who'd spent long hours after school and on the weekends writing and painstakingly editing the work of others. When I called him to talk, Sam (by now seventeen and in college) told me the project gave him not only hundreds of hours of writing practice but also the difficult but valuable experience of working with others.

"If there was an argument about an article, we would set up a discussion, and it could go for a week," he noted. The show provoked him to do academic reading in his spare time. He became intrigued by the show's recurring Egyptian symbolism and eventually wrote a four-thousand-word chapter on the subject for a book of collected essays on *Lost*. "I wound up getting pretty into it," he joked.

Now, simply turning kids loose online won't necessarily improve their formal writing. Teachers are still crucial. Kids know this: when high school students were surveyed about their writing by the Pew Research Center's Internet & American Life Project, more than half said "the writing instruction you have gotten in school" was the main reason their writing improved, and over 80 percent thought their writing would improve if teachers spent more time teaching formal writing. Nor should all writing be public. Some types of psychological and intellectual exploration require privacy, as any old-fashioned pen-and-paper journal keeper knows. But when harnessed well, digital forums can give students a reason to write more frequently, while teachers can help them to write better.

This also gives them a chance to learn digital citizenship. Heidi Siwak, an Ontario elementary school teacher, began encouraging her students to use Twitter to create ongoing class conversations. Siwak quickly realized this was a golden opportunity to learn online civics—how to respond politely and intelligently not just to friends but strangers. In a sense, she's teaching tummeling. Siwak set up several clever Twitter projects, including a daylong event where her class posted their thoughts on *Hana's Suitcase*, a book about a teenager killed in the Holocaust; as the stream got retweeted, interested strangers from Asia and Europe chimed in. Soon Siwak's class began using Twitter for other research, contacting experts on subjects like the environmental impact of Arctic exploration. "They're learning how to conduct themselves online—how to have productive conversations and exchanges," Siwak tells me.

Most attempts to use digital technology in education focus on having students learn programs: word processing programs, presentation programs, microblogging tools, search engines.

But truly clever teachers go one step further: They teach their students programming itself—how to write code. This isn't just about imparting geeky skills that will be useful on the job market. The teachers know that programming has deeper effects: For children, it becomes a philosophic act, a way of learning about learning.

This was the epiphany of Seymour Papert, an MIT mathematician and computer scientist. Back in the 1960s, Papert became interested in why so many people become "mathophobic," and he argued that it was because students are taught these subjects as dry, abstract rules. When you're exposed to high school calculus, it doesn't seem to map onto anything real. You can't easily apply the knowledge. It's like learning French in the United States, where there's nowhere to speak it. To really learn the language, you have to go to a place where it's spoken all day long, such as France, where kids pick it up in everyday life. "If we had to base our opinions on observation of how poorly children learned French in American schools, we would have to conclude that most people were incapable of mastering it," Papert wrote.

Math, geometry, and logic, Papert figured, suffer from the same problem. We assume they're hard to learn, but maybe that's because we don't live in a place where they're spoken as an everyday language. To really learn them, you can't just take lessons in school. You need inhabit a land where they're spoken every day, a "mathland."

Computer programming, Papert realized, is just such a mathland. At MIT, where Papert and his colleagues were pioneers on early computers, he'd learned that programming is more than just a technical skill. It shapes your thought and allows you to think in new ways.

Among other things, programming requires you to think logically. Computers are obedient but they're dumb; they do what you say, but *only* what you say. If you ask a human friend to get you a glass of milk, he doesn't need much more instruction. Humans implicitly know what they have to do: Go to the fridge, get a carton of milk, open it up, check that the milk is still fresh, pour it into a glass, keep the glass upright, and put the glass on a flat surface near you so it doesn't spill. But to program a computer to do the same thing, you'd have to meticulously spell out every teensy step. (If you forget to tell the program to get a glass, it'll pour the milk on the floor.) Programming requires an attention to detail and an ability to think about everything as a series of processes. What's more, computers "think" in numbers and procedures, so they're an environment where math is the native language. And because they have the brute-force ability to follow your instructions over and over, dozens or hundreds or millions of times, they never get tired. You can conduct what-if thought experiments that are impossible with pen and paper.

To give children a toehold in mathland, Papert and his colleagues created Logo, a very simple computer language. In Logo, the child controls a little turtle on-screen, issuing it commands to make it move around. The turtle draws a line wherever it goes, so it's kind of like using a computerized Etch A Sketch. To draw a square, a child would tell the turtle to go forward thirty steps, turn right ninety degrees, then do the same thing three more times. Children quickly got the hang of it, using Logo to write programs that would draw all manner of things, like houses or cars. They'd laboriously write one instruction for each step of the picture, almost the way you'd set up the dots for a connect-the-dots drawing. To draw a bird, they'd connect two quarter circles together.

But pretty soon the kids began to discover something even more fun: You could take a simple command, repeat it endlessly, and produce something unexpectedly beautiful. If you took that little bird program and adapted it to have the computer draw twenty copies of the quarter-circle while rotating it a bit each time, presto: The resulting picture looked like a *flower*. Or you could pick a seemingly simple instruction—*go forward ten steps, turn right ninety degrees, increase the number of steps by five, then repeat over and over*—and discover it produced something unexpected: a square spiral, growing eternally larger. The children began to grasp the concept of recursion, the idea that complexity emerges from repeating a simple procedure over and over. They also began to intuit the butterfly effect: how changing one tiny part of a program can radically alter the outcome. If you tweak one element in that square-spiral program, making the angle ninety-five degrees instead of ninety, surprise: The squares will shift slightly, producing a new creation, looking like a spiral galaxy. And ninety-seven degrees looks different, too.

This idea—that very small alterations can produce wildly different results—is something that many *adults* often fail to grasp, leading to massive failures in corporations, governments, teams, and families: The people at the top think that making a little change won't make much difference, but that little change spirals out of control. "Our culture," Papert wrote, "is relatively poor in models of systematic procedures." The turtle let the students think about math, and the world around them, as a series of systems: This is computational thinking.

The students also intuited many deep concepts of math on their own, merely by being able to experiment with geometry in a practical, sensual fashion. While trying to get the turtle to move in a square, they'd realize that the sum of the angles always adds up to 360 degrees. While trying to get the turtle to walk in a circle, they figured out that the trick was to have it take a tiny step, make a slight turn, and do it again and again—in essence, that a circle was nothing but a series of straight lines so short that they look like a curve. That's a foundational concept in calculus. Yet elementary school children discovered it, completely on their own, merely because they could mess around inside a world that encouraged those thought experiments.

Computational thinking isn't limited to math. Papert also had students craft poetry-generating programs. As in Mad Libs, the children would feed the program verbs, adjectives, and nouns, and the computer would combine them into lines of poetry: "MAD

WOLF HATES BECAUSE INSANE WOLF SKIPS" or "UGLY MAN LOVES BECAUSE UGLY DOG HATES." The process of trying to get the program to work lent students startling insights into language. Jenny, a thirteen-year-old girl who had previously earned only mediocre grades, came in one day and announced, "Now I know why we have nouns and verbs." She'd been taught grammar and the parts of speech for years without understanding them. But as Jenny struggled to get the program to work, she realized she had to classify words into categories—all the verbs in one bucket, all the nouns in another—or else the sentences the computer spat out wouldn't make sense. Grammar suddenly became deeply meaningful. This produced an immediate spillover effect: Jenny began getting As in her language classes. "She not only 'understood' grammar, she changed her relationship to it," Papert noted.

Indeed, teaching programming can invert one's sense of which students are good and bad at math and logic. Gary Stager, a disciple of Papert's, told me about teaching Logo Microworlds—a variant of Papert's language—to a group of young students at an international school in South Korea. Most of the kids were eagerly working on crafting programs to solve a problem. But "a little girl of about six years old, who was publicly identified as highly gifted, quickly burst into tears after realizing that the programming problem confronting her didn't have a single immediate answer," Stager noted in an e-mail. Kids who are good at traditional school—repeating rote concepts and facts on a test—can fall apart in a situation where that isn't enough. Programming rewards the experimental, curious mind.

Most important, students learn about learning itself. Computer programming is about trial and error: Few programs work the first time. Usually you've omitted an instruction or perhaps made a typo. The process of figuring out what's wrong and fixing it is exhilarating—*it works!* It's also a powerful lesson: It proves that you learn by experimenting and making mistakes, not by trying to be perfect the first time. As Papert wrote, "Many children are held back in their learning because they have a model of learning in which you have either 'got it' or 'got it wrong.' But when you learn to program a computer you almost never get it right the first time. . . . If this way of looking at intellectual products were generalized to how the larger culture thinks about knowledge and its acquisition, we all might be less intimidated by our fears of 'being wrong.'"

One of the most popular modern descendants of Logo is Scratch, a programming language for children created by MIT and distributed for free online. Using Scratch, children can create quite sophisticated games and animations. One morning I visit the class of Lou Lahana, a technology coordinator at PS/MS 188, a public school in a low-income neighborhood in New York. As the bell goes off, twenty students file into the computer lab—a modest room tucked into the corner of the building—then grab their school laptops and begin working.

Two eighth-grade kids sit next to each other, peering at one screen and puzzling over a game. They're Ruben Purrone and Esmil Sanchez, and they're working on *Ruben Invaders*—their clone of the famous arcade game *Space Invaders*. They've been fiddling with it for a few days, but it's still filled with bugs that keep it from working. For starters, when the player fires a missile at the descending aliens, it passes harmlessly through them. In technical terms, this is what's known as a "collision detection" problem.

"I don't get it—what's up with the missile?" wonders Ruben as he fires projectile after projectile.

Esmil grabs control of the trackpad and begins clicking through the game's code. It's extensive and complex: There are scores, possibly hundreds of commands in their program. "It gets hard to keep track," Esmil mutters, half to himself.

Still, they doggedly bang away at it for ten minutes until Ruben suddenly spots the problem: The missile doesn't have the right bit of code to detect the presence of the aliens. It doesn't "know" when it's made contact with them. Esmil pounces; he knows how to fix this. "We gotta set the color," he says, mousing over to the code and fiddling with its settings.

Then he hits "fire" and presto: The missiles now work. Aliens are exploding!

As I watch, the two spend the next half hour poring over the code, tweaking parameters and brainstorming solutions. (Ruben: "How do you switch the background?" Esmil: "Make the background a sprite!")

This is another side benefit of learning programming: It's a collaborative activity. Like most Scratch creators, Esmil and Ruben learned their skills not just from Lahana but also from reading guides and online forums where other kids (and adults) offer one another advice and debug one another's code. MIT hosts a site where any student can post their project for anyone to play, see, and download. Often students will download someone else's Scratch game to reverse engineer it, then tinker and upload their own version for public scrutiny. This has produced a rich culture of public thinking via code. Some Scratch players have formed their own international groups, collaborating on games remotely—with kids in Poland working alongside students in the United States, the United Kingdom, and India to create programming, music, and art. Sometimes students will discover another kid using their code, prompting debates on plagiarism that have scholarly dimensions. How much should you credit someone else if you use a bit of their code? Or a "sprite" from their game? What constitutes a contribution so creative that you can put your name on a remix?

Learning how to collaborate—particularly on a meaningful real-life project, not just a piece of schoolwork—is a crucial skill. In real life, people rarely learn and solve problems in isolation. They do it together. "Learning from others is neither new nor revolutionary; it has just been ignored by most of our educational institutions," as Douglas Thomas and John Seely Brown write in their book *A New Culture of Learning*. Yet this type of work happens very seldom in classrooms. A 2007 U.S. government survey of 737 fifth-grade classrooms found that the students spent over 90 percent of their class time working solo or listening as a class to teacher lectures, while spending almost no time—not even 5 percent of the day—working collaboratively. Obviously, not all work should be collaborative. Students need to be self-reliant, and some forms of creativity require deep solitude. (Just ask a novelist.) But current school is far too weighted toward the type of solitary work that is, for good reason, rare in daily life.

There's one last, unexpected benefit to learning programming: like public thinking, it has civic dimensions.

Young people live in a world where their daily activities are channeled by digital tools—from Facebook to photo-sharing apps to Google—yet few understand how these tools work. If you learn even a bit of programming, it is, as media theorist Douglas Rushkoff argues, like gaining X-ray vision into the digital world around you. You might begin to realize that if Facebook's much-criticized privacy settings are overly complicated, they were designed that way—and that they could just as easily have been designed a different way, less favorable to advertisers and more favorable to users. Or you'd realize that black-box electronic voting machines are inherently untrustworthy, or how digital-rights copyright protection on e-books devalues the book by transforming it from a piece of personal property to a revokable license. Much as learning law gives students tools to think about justice, learning programming gives them tools to think critically about digital life.

As Rushkoff puts it, "You gain access to the control panel of civilization. . . . Programming is the sweet spot, the high leverage point in a digital society. If we don't learn to program, we risk being programmed ourselves."

Programming games opens up new ways to learn and think. But even playing them can also do so, as Constance Steinkuehler has discovered. Steinkuehler is a professor at the University of Wisconsin who studies how and why young people play video games. This means she plays a lot herself, including games like *Lineage*—a "massively multiplayer" online world that's particularly huge in Asia. Much like the popular *World of Warcraft*, you pick a character and go on quests to beat monsters, often banding together with other players in a guild to defeat the biggest, baddest enemies—the "bosses"—which wins you the richest treasures.

Steinkuehler had joined a guild in which many players were teenage boys. They were unusually good at defeating the really hard bosses. Guild members spend a huge amount of time chatting via the in-game messaging system; despite the brief about games being "isolating," *Lineage* is as social as a pub, with kids showing up as much to talk as to play. One day Steinkuehler asked her guild members how they'd gotten so good at beating the bosses.

It turns out a group of the teenagers had built Excel spreadsheets into which they dumped all the information they'd gathered about how each boss behaved: what magical potions and weapons wounded it the most, what counterattacks the boss would employ, and how much damage each attack would cause. Like many video games, *Lineage* is quite numeric—each attack shows a number toting up the damage done. After carefully collecting all their data, the teenagers used Excel to build a mathematical model that explained how the boss worked. Then they'd use the model to predict which attacks would be most likely to beat him.

That's when it hit Steinkuehler: the kids were using the *scientific method*. They'd think of a hypothesis, like "This boss is really susceptible to fire spells." They'd collect evidence to see if the hypothesis was correct. If it wasn't, they'd improve it until their hypothesis accounted for the observed data.

"My head was spinning," she tells me. When she met up with one of the kids, she asked him, "Do you realize that what you're doing is the essence of science?"

Steinkuehler began researching conversations between players in *World of Warcraft* discussion boards and found that a shockingly high percentage of them involved "scientific" activity. Fully 86 percent of every posting was devoted to "knowledge construction," players offering hypotheses about how the game works. Meanwhile, 37 percent of the posts involved players thinking collaboratively, building on one another's ideas, and just as frequently the posts would offer counterarguments. And 58 percent displayed "systems-based reasoning"—thinking about the game as a complex environment and meditating on the rules that govern it. Here's a sample of the type of talk she found in a discussion where a player was outlining the relative powers of a mage and a priest:

> By intuition, you should notice a problem . . . but I'll give you the numbers anyways
>
> For Mindflay, SW:P, and presumpably VT [3 priest spells]:
>
> Damage = (base_spell_damage ? modifier * damage_gear) * darkness * weaving * shadowform * misery
>
> For Frostbolt [mage spell]
>
> Average Damage = (base_spell_damage ? (Modifier ? empowered frost) * damage_gear) * (1 * (1—critrate—winter's chill—empowered frost) ?
>
> (1.5 ? ice shards) * (critrate ? winter's chill ? empowered frost)) * piercing ice

mindflay = (426 ? 0.45 * dam) * 1.1 * 1.15 *
1.15 * 1.05

650.7 ? 0.687 * dam

frostbolt = (530 ? (0.814 ? 0.10) * dam) * ((1—crit—
0.10—0.05) ? (1.5 ? 0.5) * (crit ? 0.10 ? 0.05)) * 1.06

(530 ? 0.914 * dam) * ((0.85—crit) ? 2 * (crit ? 0.15)) *
1.06

0.968 * (dam ? 579.7) * (crit ? 1.15)

Please notice the 0.687 versus the 0.968. That's the
scaling factor.

This is, as the academic James Paul Gee describes it, "algebra
talk." Kids who normally couldn't care less about science were conducting university-level analysis as part of their hobby.

Steinkuehler now argues that video games are one of the best
modern conduits to teach kids about the scientific method—why
and how it works.

As she points out, many kids hate science because it's taught as a
set of facts. Indeed, that's how most adults see science: a bunch of
guys in lab coats solemnly delivering information about How the
World Works. But science isn't about facts. It's about the *quest* for
facts—the process by which we hash through confusing thickets of
ignorance. It's dynamic, argumentative, collaborative, competitive,
filled with flashes of crazy excitement and hours of drudgework,
and driven by ego: our desire to be the one who figures it out, at
least for now. Viewed this way, the scientific method is deeply relevant to everyday life, because it describes how to approach and
solve problems. But in school, students are rarely asked to actually
use the scientific method.

Games, Steinkuehler says, are an ideal native environment for
teaching the power of scientific rigor. If science seeks to uncover the

invisible rules that govern the world around us, video games are simulated worlds with invisible rule sets just waiting to be uncovered. Teachers should bring games into the classroom, she argues, so they can use them to help explain how science works.

These are fighting words. Educationally, video games are derided as a supreme waste of time and a detriment to literacy, sucking up teenagers' hours that could be devoted to reading or presumably more productive hobbies. These concerns can be valid, as I can attest; I've played video games avidly for thirty years and am painfully aware how compulsive they can become. (I had to almost completely avoid my Xbox to get this book written.) As former *Wired* editor Chris Anderson once semijoked to me, "My kids would rather play games than *breathe*." Helping students learn to moderate how much they play ought to be a crucial piece of teaching and parenting.

But Steinkuehler is also right. Games evoke modes of thinking that can be enormously valuable in education. They teach you that complex things are interesting *because* of their complexity. The trick is to learn how to use them in the right way.

Kurt Squire is figuring that out. Another professor in the University of Wisconsin's games department, he has used the game *Civilization 3* to teach low-performing kids about history, geography, and politics. In *Civilization 3*, the player picks a country and, beginning in 4000 BC, guides it through history. Players have to figure out how to devote the scarce resources of their country: Should they focus on improving agriculture? Building armies? Developing technology or artistic culture? If they're landlocked, how will they get access to water? The player interacts with other countries, which might try to invade or offer to trade. *Civ 3* is renowned for its difficulty, something like chess played with geopolitics. Squire hoped that playing would inspire the students to think about the processes that drive history, like economic development and geography and war.

He had his work cut out for him. He was invited to oversee a group of struggling students in a Boston high school. Some had absentee rates of 50 percent; all had flunked ninth grade, and one seventeen-year-old was repeating ninth grade for the third time. Most had woefully little geopolitical education. Amazingly, the school had eliminated world history as a course, because the subject wasn't part of Massachusetts's high-stakes tests. (Why teach it if the politicians weren't testing for it?) At first, the class was chaotic, the students aggressive and misbehaving. Squire tried to lecture about history to prepare the students for the game, but they had little interest. This is going to be a disaster, he thought.

But when they started playing, things changed. The students quickly discovered that *Civilization 3* was hard. Most often they'd start off frantically building armies, hoping to defend themselves against invasions. But the invading armies were always stronger, because those countries hadn't focused solely on militarization— they'd also cultivated agriculture and technology and infrastructure. The students began realizing that a country needs to thrive on many fronts; you needed guns *and* butter.

So they demanded education. They'd pester Squire with questions about history and economics and geography. One girl, Andrea, began reading encyclopedia entries and interrogating Squire about naval warfare, to try to devise a strategy that would beat her dreaded Roman neighbors. Chris, a boy struggling to play as the Iroquois, realized he needed to know more about agricultural development, so he, too, began pestering Squire for lectures ("and he wanted *details*," Squire marveled). Another student, Dwayne, applied his reading of Sun Tzu's *Art of War* to divine deeper military strategy.

The students would also collaborate, often with a sophistication that astonished the observing teachers. One day, Tristan and Tony argued about diplomacy versus war, with Tristan trying to goad Tony into building a bigger military. Tony wasn't buying it.

> TONY: Tanks? I don't need tanks.
> TRISTAN: Tony, Tony, Tony. Why don't we go by *America*'s principles? Build as many weapons as you can even though you don't need it, just in case war breaks out.
> TONY: Isn't that overkill?
> TRISTAN: What was the Cold War about? Building as many weapons as you can, just in case Russia starts something. Build enough weapons to destroy the Earth ten times over.

This political chat was coming from kids who normally were thrown out of the classroom up to 20 percent of the time for poor behavior.

What's more, they were behaving like scientists. Failure didn't kill their enthusiasm. It motivated them. When their countries collapsed, they didn't huff and complain; they wondered why, gathered more information, and ran another experiment, trying a different strategy. This was hypothesis testing, and it's the opposite of what often happens in school, where failure is punished (with a bad grade), demotivating kids. What's more, playing if-then experiments gave students an almost tactile feel for the reality of geopolitical facts that had previously seemed meaningless and dry. "I always *knew* that certain locations helped certain people," Tony said, but now "I have a better *understanding* of it." Sure, he'd been told that river valleys grew more food, that a location near the ocean helped a country grow, while also exposing it to invasion. But now he'd experienced that knowledge and prodded its dimensions. The abstract became concrete.

They also weren't gulled by the game. Squire had worried that the students would think *Civ 3* was realistic. It isn't, of course. It doesn't model the historical impact of disease, slavery, or religion. But as it turns out, the students understood that. Indeed, they'd sit around arguing about the biases built into the game—such as its bias toward conflict. War broke out far more frequently than in actual history, they discovered, because the game was trying to be dramatic.

Educationally, *Civ 3* is a rare game, insofar as its content dovetails nicely with the goal of teaching history, geography, and politics. Not many games have that property. (Though good teachers are adept at ferreting out the often-unexpected educational properties of lightweight games: One physics teacher uses *Angry Birds* to teach his students about gravity and projectile motion.) But Squire, and a new movement of "educational games" theorists, are finding that it's also possible to create games specifically to educate children, for a cost comparable to making a textbook. To try to teach the physics of charged particles to eighth graders, Squire, working with a team of designers led by Henry Jenkins, created *Supercharged!*, a game in which students assume the role of a particle flying between electric fields. The behavior of particles in these situations is paradoxical and hard to grasp; you can show a student formulas, but they don't give the student a visceral sense of what they mean.

But after playing the game, students' intuitive understanding of electrophysics bounced upward. They scored 20 percent better on a test of the concepts. When asked to draw what an electric field looks like, they created drawings more nuanced and accurate; when asked to describe it, they produced remarkably more detailed explanations. Like the *Civ 3* player who deeply got the role of geography in nation building, they got electrophysics because they'd been able to play around in a system that made it concrete and manipulable.

Video games can't do everything, of course; they're only one (new) arrow in a teacher's quiver. But their side effects can be unexpectedly powerful, including—surprisingly—their ability to improve students' ability to read.

Squire found that the *Civ 3* students began voraciously reading to try to improve their game. This makes sense, of course: When you're trying to solve an immediate problem, you're deeply motivated. (This is what the reading scholar Louise Rosenblatt calls purpose-driven "efferent" reading, in contrast to the "aesthetic" pleasures of losing yourself in a novel.) School rarely motivates students to read in this urgent, engaged way because school rarely offers children any problems they find particularly urgent. Games, in contrast, are designed to provide you with problems so urgent and tantalizing you can't stop thinking about them.

Steinkuehler also noticed that players were reading to improve their play—everything from discussion forums to wikis to walkthroughs. She wondered: Was this reading actually challenging? Indeed it was. She took a group of boys aged twelve to eighteen who played *World of Warcaft* and who all read, at best, at grade level, which is pretty much in keeping with the lackluster reading performance of young men on average. But then Steinkuehler asked them to pick a problem they were currently grappling with in *World of Warcraft*. The researchers selected a text for each boy to read that related to the game problem he was trying to solve, specifically picking texts that were lexically complicated and challenging. Yet the students were able to read those texts with ease: The nonstruggling readers tackled texts that were 3.5 grade levels higher than their typical abilities, and boys classified as struggling improved even more, reading texts fully 6.2 grade levels higher.

What happened? Why did their ability to read suddenly increase so dramatically? Because they were interested in the subject, making them willing to ponder and deduce the meaning of more complicated language and unfamiliar, even academic, terms.

As Steinkuehler writes, "Interest does matter": The kids were trying to solve a problem they cared about. Games can provide a pathway for teachers to reveal what students are capable of. And as Squire has shown, they can hook students into reading deeply and excitedly in everything from history to economics.

Ever since Marc Prensky coined the term "digital natives," we've been told that young people have an innate edge in using the technology. They're comfortable with it; they get it, effortlessly, in a way older people don't. But this, alas, isn't really true.

That's what Bing Pan, a business professor at the College of Charleston, discovered in a clever experiment. He wanted to test students' facility on an omnipresent digital skill: How adept are they at using Google? So Pan asked a group of them to use the search engine to answer several questions. As you might expect, the students favored the top few links that Google returned.

Then Pan artificially inverted the results on the first page Google returned, putting the tenth result in the number-one slot and so on. More often than not, the students took the bait and again favored the first links—even though they'd been put there falsely. As Pan realized, the students were not actively evaluating the actual relevance of the results. They just trusted the machine.

Other studies have found similarly dismal results. A study of 102 Northwestern University undergraduates found that none ever bothered to check authors' credentials on a Web site. Another found that more than a third of college students were unaware that search engines include paid-for links in their results. These were students who'd been using the Internet, on average, for *seven years*. In other words, digital natives might feel like they've mastered their tools, but that doesn't mean they truly understand how they work. This ignorance is intellectually crippling, because the results on Google (and all search engines) are prone to all manner of artificial gaming and corporate juking. The upside of public thinking—that anyone can publish—is matched by its perfectly inverted downside, which is that anyone can publish, leaving the online environment devoid of the marks of hierarchical authority on which students for centuries have relied. When I was in elementary school in the 1970s, the biggest resource we had was a couple of sets of encyclopedias. We weren't asked to judge whether they were accurate or not; the school system and librarians took care of verifying that. (Certainly, the encyclopedias had their own deficits; they became quickly out of date and were, compared to today's online resources, woefully narrow.) In the 1950s social critics pondered "Why Johnny Can't Read." Now they should ponder "Why Johnny Can't Search."

Whose fault is it? Not the students'. If they're unable to navigate online information, it's because, rather amazingly, they're almost never taught search literacy in schools. It ought to be a core part of what kids learn in school (and new common core standards in the United States are beginning to emphasize it), but for years it was barely touched upon. This is surpassingly ironic, because teaching search literacy is a golden opportunity to teach critical thinking: What am I being told? What motivations does this person have for telling me this? Does the information match other things I know? Is it even checkable or is it speculation? These are the skills that adept adults deploy, often unconsciously, when they search for information online.

The thing is, it's quite possible to train kids in search literacy. Indeed, librarians worldwide are the heroes in this story. They're frantically working on teaching those skills, picking up the ball that the curriculum has thus far dropped.

Consider the efforts of Frances Harris, librarian at the magnet

University Laboratory High School in Urbana, Illinois. Harris takes eighth and ninth graders and puts them through a search boot camp, showing them how use advanced settings and alternative search engines that feature curated content and don't track their users. She steers them away from raw Google searches, pushing them toward academic and news databases, too. And crucially, she trains them to check the credibility of sites they find—whether it was written by an academic, an advocacy group, or a hobbyist. She trains them not to be fooled by professional-looking design; plenty of corporate sites look flashy while peddling self-serving infojunk.

It works. Within a few weeks, the students can pass several adroit tests. For example, they begin to detect when Google is serving up pure crap (like a sham Martin Luther King site that's actually run by white supremacists) or, more commonly, the gray-area stuff like content farms, Web firms that flood the Internet with dull, minimally informative articles just to sell ads.

"It's not the outright lies that you have to teach them to watch out for," Harris tells me. "It's just this vast sea of *mediocre* stuff. But I see them start to get really paranoid. They keep on asking, 'Wait, wait, is this a content farm?' And this is what you want. Most people in their lives aren't going to be writing term papers, but they're going to be looking for information their whole lives."

Crap detection, to use Howard Rheingold's phrase, isn't easy. Among other things, it's easier to do if you already know about the world. For instance, Harris found that students had difficulty distinguishing a left-wing parody of the World Trade Organization's Web site from the real WTO site. Why? Because you need to understand why someone would want to parody the WTO in the first place—knowledge the average eighth grader does not yet possess. In other words, Google makes broad-based knowledge of the world *more* important, not less.

Many pundits, myself included, argue that kids should spend more time with books, since they're a time-tested vehicle for deep thought. Librarians do push kids toward books—but, brilliantly, they extend the critical thinking there, too. As Buffy Hamilton, a librarian at Creekview High School in Canton, Georgia, points outs, books also frequently contain flat-out factual errors, flaccid arguments, and loopy biases. And unlike digital documents, these errors can't easily be fixed or called out in comment threads. "We can all think of situations where there are factual problems in printed books. Or that academic paper that was peer reviewed but then, whoops, it turns out to be wrong," she tells me. It's the dirty secret of the book world that virtually no publishers vet facts before they're printed. "Let's not bash on the Web and Wikipedia. Let's check *everything*," Hamilton says.

Training students to use a new tool for thought is never easy, and historically it's been done rarely and unevenly. Consider that even an old-fashioned research library is bewildering if you're not shown how to use it, and a lot of students never are. When I arrived at the University of Toronto in 1987, I—being an apple-polishing nerd—signed up for an orientation session for the university's sprawling Robarts Research Library. Only a handful of students showed up. Assuming the other sessions were as sparsely attended, I figured that barely a sliver of the four thousand new students each year were actually trained to use the library. It showed: My professors complained that students barely sampled any secondary sources and never consulted academic journals. When CD-ROM collections of journals emerged in the 1980s, librarians complained that even *professors* weren't bothering to learn to use them correctly. In a prefiguring of today's Google illiteracy, they'd type in a single word for a search, print out the top five hundred results regardless of whether they were relevant—then happily walk away, feeling like assiduous researchers simply because they had generated a massive pile of *something*. (They were the "inept and satisfied end user," as University of Alabama librarian Scott Plutchak called them.)

Our new tools are powerful, but only if we're taught to use them. The lucky part is that students are in an environment—school— where society has a shot at instilling its intellectual values, skills, and culture. It's up to us to make sure students are taught to use these new tools, instead of being used by them.

Ambient Awareness_

Who cares what you ate for breakfast?

That question has become a cliché of Internet criticism, the go-to response to social networking sites like Facebook and Twitter. And there is, it's true, something off-putting about the world's newest literary form, the bite-size "status statement," at least at first glance. There's the 140-character tweet about a celebrity or the Facebook link to a gushy news story you just read or a picture of your cat filtered to look like an acid flashback. Then there's the "like"—a single flip of a social bit. When you consider the oceanic volume of this stuff, you might well conclude that it's further proof that the Internet has shriveled our attention spans and strip-mined human intimacy. Why do we post so many teensy utterances? And why do we so eagerly devour them?

Ben Haley, a technical support specialist in Seattle, pondered this puzzle when, at the urging of a friend, he first signed up for Twitter in 2007. Like many, he couldn't figure out why anyone would care about such brief messages. One friend tweeted about how she was becoming sick. Another posted links to random stories he was reading. Yet another—straight out of that playbook for vapid updates—would describe her lunch, every single day. Each tweet was so brief as to be virtually meaningless.

But as the months went by, something changed. By following his friends' updates, Haley began to sense the rhythms of their lives. He developed a mental map of what they were doing and even thinking. He could tell when his friend was recovering from her illness. He could track his friend's obsessions by seeing (and sometimes reading) articles he was linking to. Even the litany of sandwiches became spellbinding, a glimpse into the cadence of his friend's life, wryly humorous and even poignant in their detail.

The flow began to seem like "a type of ESP," Haley told me, an invisible dimension of information floating above everyday life. "It's like I can distantly read everyone's mind," he added. "I love that. I feel like I'm getting to something raw about my friends. It's like I've got this heads-up display for them." It also led to more real-life contact: When one member of Haley's group broadcasted his plans to go to a bar, the others would see it, and some would drop by—a type of ad hoc self-organizing that has become common in young people's lives. And Haley also noticed that when he did socialize face-to-face, the conversation was subtly altered. He and his friends didn't need to ask, "So, what have you been up to?" because they already knew. Instead, they'd begin discussing something one of the friends posted that afternoon, as if picking up a conversation in the middle.

Social scientists have a phrase for this type of ESP: "ambient awareness." Ambient awareness is, they say, almost like being in the same room as someone and picking up on his mood and thoughts by the stray signals he gives off. You create a picture of someone else's internal state gradually, almost unconsciously, by assembling many small observations.

Mizuko Ito, a cultural anthropologist, first noticed this effect more than ten years ago while studying text messaging in Japan. Ito talked to young Japanese couples, all of whom lived in separate apartments (and in one case, separate cities). She found that the couples would trade short text messages all day and night to establish a sense of connection. They'd ping each other with tidbits like "I guess I'll take a bath now," "Just bought a pair of shoes," or "The episode today sucked today, didn't it?" One young businessman jokingly described his text messages to his girlfriend as "mutterings" and the responses he got as "mutterings in reply."

The result, Ito discovered, was that they felt uncannily proximal to each other. Their sense of co-presence, Ito writes, was "similar to the kind of awareness of another one would have when physically co-located . . . a way of entering somebody's virtual peripheral vision." Or as she explains to me, "It's like you're in the room and you just sort of share a sigh or a facial expression."

This is the paradox of status updates. Each little update—each individual bit of social information—is, on its own, pretty insignificant, even mundane. But taken together over time, the snippets coalesce into a surprisingly sophisticated portrait of your friends' inner lives, like dots forming into a pointillist painting. "It's an aggregate phenomenon," Marc Davis, a partner architect at Microsoft, tells me. "No message is the single-most-important message."

Before modern technology, this type of awareness wasn't possible. No friend would bother to call you up daily and detail the sandwiches she was eating, the articles she'd read, or the miles she'd walked today; indeed, if she did you'd have have found it annoying and intrusive. But ambient tools weave this knowledge into a tapestry you can glance at, which makes the picture both more complete and more inviting. Unlike a series of phone calls, it's optional—so, as Davis points out, it invites your attention, rather than demanding it. (Or, as he adds, it's like the difference between a friend in the 1970s forcing you to sit through their agonizingly long vacation slide show and the friend posting the photos online so you can riffle through whichever and whenever you'd like.)

Right now, if you concentrate for a few seconds, you can probably conjure up a recent map of the doings—even the *thinkings*—of the people you follow. I physically talk with my college friend Bret only once a year, when I visit Toronto at Christmas. But I nonetheless know what he's up to right now, including: After reading *The Intuitionist*, a novel about elevator repair, he became obsessed with hunting down vintage-elevator videos on YouTube; that he's doing a lot of weird cooking experiments, including making high-end processed cheese; that he recently rediscovered posters of his university band and started playing music again; and that watching the Mars *Curiosity* rover landing restoked his Canadian appreciation for collective government action. There are professional colleagues I follow who "thoughtcast" their work—like the sociologist Tricia Wang, who spent seventeen months traveling through the impoverished provinces of China and posting pictures on Instagram (trains jammed with migrant workers, hungry schoolchildren receiving their first free school lunch). And like many people, I follow interesting strangers—great curators, provocative thinkers, or simply famous folks whose posts intrigue me.

Of course, it's my personal interest in my friends (and these strangers) that makes their feeds meaningful. If *you* were to glance at the people I follow, you'd just see disjointed info. You wouldn't have developed the big picture that makes each small utterance interesting. Vice versa is also true: If I look at yours, I just see noise. This is precisely why critics of social media can so easily point to any individual status update and proclaim that it's a snippet of meaningless tripe. Without context, it can certainly look that way.

But ambient awareness is all about slowly amassing an enormous, detailed context. Follow someone's ambient signals for a day and it seems like trivia. In a week it seems like a short story. In six months, a novel.

This isn't just idle interest in others' doings, though. Ambient awareness also endows us with new, sometimes startling abilities. When groups of people "think aloud" in this lightweight fashion, they can perform astonishing acts of collaborative cognition. Scientists have known about this for years because they've seen it happen offline, in the physical world.

Back in the early 1990s, for example, the British social scientists Christian Heath and Paul Luff began studying the coordination of staff in the control rooms of the complex London underground system. If one train becomes delayed or drivers aren't available, the controllers have to quickly "reform" the system, deciding which trains need to be taken out of service, which should be rescheduled, and which drivers should be assigned to which trains. "Reforming" is a challenge because different staff members need to be on the same page and aware of each other's actions—but they don't always have enough time to confer about who's doing what.

To solve this problem, the control staff developed a clever technique: They would talk aloud. Each time they altered the system, they'd enunciate what they were doing. They weren't talking *to* anyone in particular. They were doing ambient broadcasting—speaking into the air, for anyone to hear. This sort of "self talk," Heath and Luff noted, helped the group very rapidly establish a common awareness of what was going on. It could even establish the thinking that was going on: By talking out loud, the controller "renders visible to his colleagues the course of reasoning involved in making particular changes," as Heath and Luff wrote. With this level of group awareness, the staffers could perform clever feats of collective problem solving. In a sense, they were doing status updates before status updates existed. And subway-control workers aren't the only ones who use this technique, either: Huff and other academics have documented the talking-aloud-to-the-room strategy in surgical teams, newsrooms, airport luggage-control rooms, and financial-industry teams.

Today, this sort of group awareness happens constantly online. I think of it as a form of proprioception, our body's awareness of where its limbs are. That subliminal sense of orientation is crucial for coordination: It keeps us from accidentally bumping into objects and makes possible extraordinary feats of balance and dexterity. When you're able to pass a baseball from your left hand to your right hand behind your back, that's proprioception. When groups of people—friends, family, workmates—keep in lightweight online contact, it gives us *social* proprioception: a group's sense of itself.

Social life is now filled, in a manner at once banal and remarkable, with just this type of self-organization. When my wife went on a work trip to Los Angeles, she originally didn't plan to do any socializing. But when she wrote a Facebook status statement about watching a Demi Moore movie in her hotel room, she began getting pinged by old friends whom she had forgotten lived in the city. (She wound up with invitations to dinner and a concert.) At professional conferences, attendees use everything from tweets to pictures and check-ins to develop a communal sense of what's going on and what interesting presentations each person might have missed, providing fodder for conversations when they all meet up face-to-face at dinner.

These are simple, quotidian examples, but serious ones abound, too. Among political activists, ambient contact has proved particularly powerful in moments of crisis. In November 2011, the Egyptian American journalist Mona Eltahawy was arrested by the Egyptian security forces during a protest and fired off the single tweet: "Beaten arrested in interior ministry." Friends and followers worldwide, accustomed to glancing at Eltahawy's utterances online, instantly learned of the arrest and began talking about it. One was Zeynep Tufekci, an assistant professor who studies technology and society at the University of North Carolina at Chapel Hill; she contacted NPR's Andy Carvin, a friend who tweets about the Middle East, and the two set up a hashtag—#FreeMona—to coordinate how to help Eltahawy. A mere twenty minutes later, so many supporters were discussing how to help that #FreeMona became a worldwide trending topic. Hundreds more sprang into action: Sarah Badr, a designer, called the U.S. embassy and tweeted her conversation alerting them; Anne-Marie Slaughter, a former director of policy planning at the U.S. State Department, contacted her former employer; and Egyptians began dialing Egyptian media and al-Jazeera to encourage them to cover Eltahawy's case. It's not easy to pressure the Egyptian military, but the coordinated burst of activity worked. The next day, the military released Eltahawy, whose left arm and right hand had been broken during her ordeal.

Ambient awareness on a global scale is strange and new. But it works so well because it taps into older social skills, including our ability to "read" other people. Indeed, when you're in regular enough digital contact, even *silence* becomes a readable signal—much as you can sense someone's mood shift at a dinner table if she clams up. For example, Lisa Hickey, a journalist I know, one day began to subconsciously sense that "something was up" with an old friend. "When I finally contacted her, it turned out that she'd suddenly gone into the hospital for a life-threatening emergency," Hickey says. At that point Hickey realized, post hoc, where her intuition had come from: Her friend, normally fairly chatty on Facebook, hadn't posted anything for several days and also hadn't posted any explanation for why she'd gone dark. It was the sudden quiet that had alerted Hickey. "And I realized, Oh, this is the invisible ESP."

Indeed, our online and offline people-reading skills appear to be quite closely linked. The personality psychologist Sam Gosling has conducted clever experiments that illustrate how our real-life, physical possessions reveal our personalities. In one, subjects gave Gosling permission to bring strangers into their houses while they were away. After only a few minutes examining their bedrooms, the strangers could produce eerily accurate descriptions of the subjects' personalities. Gosling wondered whether Facebook pages would be equally revealing, so he repeated the experiment online, arranging to have strangers inspect the Facebook pages of experimental subjects. Sure enough, they were able to accurately describe those personalities, too. The ambient signals given off by status updates and streams of photos can be as powerful as those from real-life objects.

Indeed, they're sometimes *more* revelatory. One of the hilarities of ambient life is discovering how much weirder people are than you thought, even those you believed you knew well. Jack Dorsey, the creator of Twitter, once told me that when his parents began using the service, their updates revealed sides he'd never known. "They like to party. They're big fans of going out and drinking!" he said. "And they like to *cuss*. And I learned some of their eating habits, and just a lot of stuff I didn't consider. Text is very freeing, especially short bits of text. It allows you to abstract yourself and reflect in a different way. So my mom just . . . writes in a way I never heard her speak."

Ambient contact doesn't just reshape life online. It's also altering the way we interact physically.

You can spy this in the work patterns of younger employees in the United States. As they've moved into the workforce, they've developed intriguingly different ways of relating to the office and their colleagues than their parents' generation, the boomers.

"Baby boomers grew up in a time when you had to go to the office because you had a phone connected by a wire, and you had to go there to use that phone," as Michael O'Neill tells me. He's a researcher who, working for the office-design firm Knoll, has polled over twenty thousand workers of all ages. He found that boomers regarded being in the office as crucial to work. This makes sense; in the past, the office was, after all, the only place colleagues could quickly establish group awareness and make decisions. So they grew accustomed to the formal sit-down meeting as a way to establish group awareness and make decisions. "The unit of work for the baby boomer is the formal meeting—'Bring on the coffee, and we're going to sit here for three hours,'" O'Neill adds. But younger workers were completely different. They found traditional meetings vaguely confrontational and far preferred short, informal gatherings. Why? Because they were more accustomed to staying in touch ambiently and sharing information online, accomplishing virtually the tasks that boomers grew up doing physically. Plus, the younger workers had the intuition—which, frankly, most older workers would agree with—that most meetings are a fantastic waste of time. When they meet with colleagues or clients, they prefer to do it in a café, in clusters small enough—no more than two or three people— that a serious, deep conversation can take place, blended with social interaction, of a sort that is impossible in the classic fifteen-person, all-hands-on-deck conclave. Many young entrepreneurs I interview these days put off getting an office for a surprisingly long time, regarding it as a drag on resources and even productivity; instead, their employees all work from home or cafés or coworking spaces around town (or around the world), keeping in constant lightweight contact via tools like chat and Skype. "I think the biggest office supplier right now, in terms of being a landlord, is Starbucks," as Scott Annan once joked to me; a Canadian entrepreneur, Annan ran his software firm Mercury Grove with a staff of twelve, for more than five years without an office, and now runs the startup Accel.io.

In a similar fashion, ambient contact is beginning to kill off that earlier technology of contact—the phone call. According to Nielsen data released in 2010, the number of mobile calls per person peaked in 2007 and then dropped steadily. And those calls are getting shorter: in 2005 they were three minutes on average, and by 2010 they were half that length. Throughout most of the twentieth century, of course, the phone call was the *only* way to quickly learn what your remote friends, family, and colleagues were doing. But our reliance on the phone disguised the fact that it was actually a dreadful tool for keeping tabs. It was inherently interruptive: most of the time when you called someone, you weren't sure if they were free to talk and the callee wasn't sure why you were calling—so you had to open up Schrödinger's box every time, having a conversation just to figure out if it's okay to have a conversation. Once ambient awareness became available, those "What are you doing?" calls began to vanish. The phone calls that remain are ones that are pre-scheduled (presaged by the "you free to talk now?" text) and thus can involve richer, lengthy conversations.

Used correctly, social proprioception can make one's work life less frazzled by reducing the constant stream of interruptions. A group that's connected in an ambient fashion can—counterintuitively—spend *less* time on communication, particularly the writing and reading of endless e-mail. And e-mail has indeed become one of the banes of corporate existence. Office workers spend an estimated 28 percent of the workweek writing and reading the stuff, a load that's growing by 7 percent a year.

Five years ago in 2008, Luis Suarez hit the e-mail wall. A long-time IBM employee based in the Canary Islands, Suarez tells me that he found himself spending more than two hours a day merely on e-mail. When he analyzed his inbox, he noticed a couple of trends. A big chunk of the e-mails he got were short queries from colleagues trying to figure out his status—questions like "Where are you?" or "How's the project going?" Another chunk were questions from colleagues that could more easily be answered by someone else. Was there any way to get people to stop sending him these messages?

He decided there was: first, by making his work life an open book.

Suarez began constantly broadcasting what he was doing, posting several times an hour to various sites, including Twitter and IBM's internal social network. He talked about what he was working on, where he was, and who he was with. Soon, his colleagues began sending him fewer status-seeking e-mails, because they already had the answer; a glance at his feed was faster and more complete than a query. Second, Suarez told his colleagues that if they had a question for him, not to e-mail it but it post it openly on IBM's internal social network. This produced some interesting results. When a question was posted, often someone else would answer it more quickly—and more accurately—than Suarez could. And when Suarez did jump in, the answer would remain in public permanently for everyone to see, building up into a huge knowledge base, an external file of Suarez's smarts that any IBM staffer could tap into.

Because he was living more publicly at work, Suarez's e-mail load began to shrink. Within a few months, it was down by half; within a year, it was down by an incredible 85 percent. Today, he fields barely two e-mails per day, mostly sensitive communications. "The more you open up," he says to me, "the more you reduce the need for people to send you messages."

Some research backs this up: a survey of 912 white-collar workers found that those who made themselves ambiently available through instant messaging reported less interruption than those who relied more on other forms of communication. You might expect instant messaging to make things worse, with colleagues pinging you all the time. But it had the opposite effect: It became easier to quickly figure out if someone was free ("you got a second?") and easier to tell someone to wait if you were busy. The negotiations were more like the casual check-ins you get between people sitting and working at a table together.

At first Suarez seemed like a crazy outlier: "People thought I was nuts." He laughs. Now companies are trying to follow his lead; French tech-services giant Atos, with seventy-four thousand employees, aims to be e-mail free by 2014, and for some unionized employees, Volkswagen turns off e-mail to their corporate Black-Berry phones overnight so they won't be bothered by work messages in their off-hours. This doesn't have to be an all-or-nothing prospect, either. Even shifting a small amount of your thinking from e-mail to social networks can have big effects. A study by the McKinsey Global Institute estimated that if employees at firms used more ambient media, it could reduce e-mail by 25 percent and reduce the time searching for information by 35 percent.

The rise of short-form social media hasn't gone over well with everyone.

Indeed, social critics have been withering in their disdain. The status-update world, they say, is a parade of narcissism—vapid, inward looking, and idiotic. Twitter and Facebook, in particular, have been lightning rods for these judgments. "Who really cares what I am doing, every hour of the day?" wondered Alex Beam, a *Boston Globe* columnist, in an early essay pondering Twitter. "Even I don't care . . ." As this complaint usually goes, ambient signals are far too brief to convey any serious meaning or conversation—indeed, they train a young generation to think only in bite-sized, nonlinear, agrammatic chunks of LOLspeak. Worse, they channel our endless appetite for self-regard, turning everyone into the star of their own private reality show. And they push into hypermetabolic overdrive the modern culture of personal branding, encouraging everyone to regard their life as a corporate endeavor, racking up massive tallies of followers and "friends"—friends who, of course, bear absolutely no relation to actual friendship—merely in the service of having a megaphone into which the Dunciadic masses can overshare nothing of any discernible value.

Twitter, as the editors of the literary magazine *n+1* despaired, can open a portal onto "the narcissistical sublime." They continued: "What tweet is that, flashing, subliminally, behind the others? In exactly 140 characters: 'I need to be noticed so badly that I can't pay attention to you except inasmuch as it calls attention to me. I know for you it's the same.' In this way, a huge crowd of people—40 percent more users since last year—devalue one another through mutual self-importance." Bill Keller of *The New York Times* denounced Twitter as "the enemy of contemplation." And don't even get these guys—and they're very often guys—started on sites like Pinterest, where women (and they're very often women) broadcast dioramas of scenes and pictures they find aesthetically lovely. "Dreck," as one of my high-tech writer colleagues sniffed.

It's undoubtedly true that plenty of people online can seem self-involved, and that these tools exacerbate it. The tools are, after all, designed to *encourage* self-disclosure. Social critics, including Christopher Lasch in his famous book *The Culture of Narcissism*, have argued for decades that self-regard is on the rise. Lasch pinned a good chunk of the blame on the erosion of structures that, as he saw it, historically encouraged people to focus outside themselves, such as religion, family, meaningful work, and class. Deprived of traditional roles, Lasch argued, American identity in the twentieth century became fragile, requiring incessant attention from others for validation. In recent years, psychologists like Jean Twenge have argued that there's empirical proof of an epidemic of narcissism, though the data isn't necessarily that clear (other evidence suggests young people's personalities haven't changed much over several decades). Still, even if you don't grant the full force of these sweeping proclamations, it's hard to disagree that much of what happens in the status-update universe is pretty banal. As Clay Shirky has pointed out, this is an inevitable side effect of mass publishing. You can't uncork so much expression without seeing a plunge in the quality of the average utterance.

But I suspect that what's really going on is not terribly causal. I doubt the ambient broadcasting universe is *making* people more trivial. What it's doing is revealing how trivial we've been all along, because it's making conversation suddenly visible.

We might flatter ourselves by pretending that before Facebook and texting and "thoughtcasting," our conversation was uplifting and erudite—focused on higher culture and civic life, studiously avoiding self-regard and gossip. But research suggests that's a myth. When British anthropologist Robin Dunbar studied everyday face-to-face conversation in Britain, he found that people spent fully 65 percent of the time talking about themselves and other people: What you could politely call social experiences, or less politely, gossip. (The only big difference, at least among younger people, was that men were far more focused on themselves. When talking about social experiences, women spent two thirds of their time talking about other people's social experiences; men spent the same amount of time talking about their own.) Dunbar doesn't think our deep interest in the doings of others is a bad thing, mind you. His hypothesis is that language, one of our signal intellectual gifts, evolved precisely to allow us to socially "groom" one another, in the way that primates groom one another physically. It is the glue of society. When critics freak out about the triviality of online mobs, they are probably not listening carefully to their own daily talk. Or more charitably, before the Internet came along, they lived in a bubble where they could at least hope that the daily talk of the hoi polloi was uplifting, because they had no way of eavesdropping on it. Now they do. Digital tools are making the previously invisible visible, which is always unsettling.

In fact, the historical pattern here is steady: Each new tool for communication has provoked panic that society will devolve into silly chatter. As Tom Standage notes, we saw this during the rise of coffee shop culture in the seventeenth century. Today, intellectuals often wax nostalgic about early European coffee shops, where students and writers and businessmen would gather to read newspapers and debate weighty issues. But at the time, authorities hated them. They worried that café going would destroy young students' habits of concentration; the café gossip was seducing them away from their studies. "The scholars are so greedy after news (which is none of their business) that they neglect all for it, and it is become very rare for any of them to go directly to his chamber after prayers without first doing his suit at the coffee-house, which is a vast loss of time grown out of a pure novelty," as one Cambridge observer argued. "For who can apply close to a subject with his head full of the din of a coffee-house?" In the 1673 pamphlet "The Character of a Coffee-House," the author savages the social hangout for being "an exchange, where haberdashers of political small-wares meet, and mutually abuse each other, and the publick, with bottomless stories, and headless notions; the rendezvous of idle pamphlets, and persons more idly employed to read them. . . . The room stinks of tobacco worse than hell of brimstone, and is as full of smoke as their heads that frequent it."

A little over a century later, the same panic emerged with the next New Thing: the mass-market novel. The growth of penny dreadfuls—particularly romances and gothic horror novels—could not possibly be good for one's mind, could it? Not according to the 1835 essay "Devouring Books," which fretted over the surge of "novels and light periodicals" from "the teeming press." "Perpetual reading," the authors proclaimed, "inevitably operates to exclude thought, and in the youthful mind to stint the opening mental faculties, by favoring unequal development. It is apt either to exclude social enjoyment, or render the conversation frivolous and unimportant; for to make any useful reflections, while the mind is on the gallop, is nearly out of the question; and if no useful reflections are made during the hours of reading, they cannot of course be retailed in the social circle." Women, being so delicate, were of course at most danger. As Jonathan Swift put it (possibly in parody; it's hard to tell with Swift), all that reading was "apt to turn a woman's brain," rendering her "vain, conceited, and pretending," and causing her to "despise her husband, and grow fond of every coxcomb who pretends to any knowledge in books." With each wave of panic, critics turned to the cutting-edge science of the day to explain the devolution of our brains. Today social critics point to brain scans as objective proof that our brains are being rewired by media; in the nineteenth century, they pointed to the nervous system. U.S. neurologist George Miller Beard diagnosed America's white-collar population as suffering from neurasthenia. The disorder was, he argued, a depletion of the nervous system by its encounters with the unnatural forces of modern civilization, most particularly "steam power," "the telegraph," "the periodical press," and "the sciences." (Beard also noted a danger specific to the nerves of the fairer sex: an increase in "the mental activity of women.")

Soon, the same panic emerged in response to the telephone. It would, critics predicted, atomize society into a blasted landscape of pasty, sun-averse morlocks unable to socialize face-to-face, because they'd be out of practice. Who would bother to leave the house, when they could simply call someone? Worse, it would degrade human interaction into a rambling exchange of trivialities, as Mark Twain suggested in his 1880 satirical sketch "A Telephonic Conversation." Meanwhile, mavens of etiquette fretted that the telephone would coarsen our manners, because the predominant greeting— "hello!"—derived from the shout of "halloo," a bellow used to summon hounds to the hunt. (Americans fought about the propriety of "hello" until the 1940s.) Today, of course, everyday use has so domesticated the technology that nostalgics now regard the telephone as an emotionally vibrant form of communication that the Internet is tragically killing off.

This isn't to say that technology changes nothing for the worse or that these criticisms weren't partly correct. "Technology is neither good nor bad; nor is it neutral," as the technology thinker Melvin Kranzberg pointed out. It's why each tool needs to be carefully scrutinized on its own merits. The stuff we find merely annoying about a new technology in the short run is often off base, and we can miss the subtler ways it alters patterns of thought in the long run. Personally, I suspect a bigger danger of modern ambient contact is that it reinforces the recency effect—making us feel that information arriving *right now* is more important than events that happened yesterday, last month, or last century. We've long struggled, as a culture, to pay attention to the past. It's probably harder now than ever.

So no, I don't think new media are bringing on a social dystopia. But they're not bringing on a social utopia, either. That's an equally metronomic pattern: For every technological doomsayer in history there was a rival prophet of glee, promising that new communications would erase our problems. War and conflict, in particular: Back in the nineteenth century, the telegraph was supposed to put us in such intimate global contact that killing a foreigner would seem as barbaric as killing your neighbor. "This binds together by a vital cord all the nations of the earth," gushed the authors of the 1858 book *The Story of the Telegraph*. Decades later, the airplane would help humanity transcend itself: the "aerial" man, wrote the feminist thinker Charlotte Perkins Gilman in 1907, "cannot think of himself further as a worm in the dust, but as a butterfly, psyche, the risen soul." Then it was TV that would bring about the end of war; then the Internet and mobile phones.

The comedian and writer Heather Gold, one of the cocreators of the concept of tummeling, once told me that social media is unsettling to many because it *feminizes* culture. All this "liking," this replying, these bits of conversational grooming—"phatic" gestures, as linguists would call them, which constitute a significant chunk of our ambient signals—are precisely the sorts of communiqués at which women are traditionally urged to excel. "You go online and all these type-A, alpha-male business guys are acting like thirteen-year-old girls, sending little smilies to each other publicly and going *hey, happy birthday!*" she told me. Obviously, these are crude categories; many men have superb social skills, many women have terrible ones, and as feminists have long noted, the relegating of women to "social" jobs is part of how they've been sealed out of decision-making roles for millennia. But this shift toward a world that rewards social skills is real.

To put it another way, reaping the *cognitive* benefits of the Internet often requires *social* work. This distresses anyone for whom social work is a chore or seems beneath them.

Technically, I owe my new house to ambient awareness. Back in late 2009, my second child was born and our family had outgrown our apartment. We decided to try to buy a house in Brooklyn, NY, so we started taking weekend trips out to neighborhoods that seemed suitable. We'd been at it for many months, not finding anything that worked. Then one day during the hunt, Emily wound up in the subway station at the southern end of Brooklyn's Prospect Park. She looked at her phone and noticed that there was a WiFi signal inside the station, deep underground. So she posted a status update on Facebook: "There's WiFi in the 15th Street subway stop. Who knew?" As was typical, a few of Emily's close friends commented in reply—they were surprised, too.

But then something fortunate happened. A woman named Anne also noticed the status update. Now, Anne wasn't a close friend of Emily's; in fact, Emily hadn't seen her in years—she was married to Brad, a man Emily had met fifteen years previously when they'd both lived in Atlanta. Brad and Anne had lived in Brooklyn, but once they moved to the Midwest, Emily was aware of their status only on Facebook, in that sort of jumbly, promiscuous oh-hi-there-let's-stay-in-touch way.

Anne happened to glance at the subway comment and asked, in the thread, "What are you doing in that neighborhood?" Emily explained that we were looking to buy a house. Then Anne pointed out that hey, coincidentally, *she* had some friends who were just about to put their house on the market in that area. Next week, in fact!

Quickly the dominoes began to topple. Anne put us in contact with the owners. We discovered that the house was precisely what we wanted. And because we'd heard about it through word of mouth, we were able to put in an early bid. A few months later, we moved in.

This entire stroke of luck hinged upon what sociologists call a weak tie. In a pre-social-network world, there's almost zero chance that Emily would have been in regular contact with the wife of a friend from years ago. One's social circles rarely include that sort of distant connection. It's even less likely that a distant link would be exposed to your stray observations, like "Wow, there's WiFi in this subway station." In a world of status updates, tangential, seemingly minor ties become part of your social fabric. And they can bring in some extremely useful information.

In 1973, sociologist Mark Granovetter gave a name to this powerful process: "The Strength of Weak Ties." Granovetter had spent time researching the ways in which people found new jobs. After surveying hundreds of job finders, he discovered there were three main strategies. You could use formal means, by responding to job advertisements. You could try direct application, cold-calling companies to see if they had any openings. Or you could harness personal contacts: You'd ping everyone in your social network, asking them if they'd heard of any recent jobs.

When Granovetter analyzed the data, two things leaped out. First, jobs that people heard about via personal contacts were best of all. These jobs were more likely to have high salaries than jobs found through formal means or direct application, and they were more likely to be fulfilling: more than 54 percent of the people who'd heard of their job through personal contacts were "very satisfied" with the new job, compared to 52.8 percent who'd used direct application and only 30 percent of those who'd used formal methods.

But the second finding was even more intriguing: When people got these word-of-mouth jobs, they most often came via a weak tie. Almost 28 percent of the people heard of their job from someone they saw once a year or less. Another 55.6 percent heard of their job from someone they saw "more than once a year but less than twice a week." Only a minority were told of the job by a "strong tie," someone whom they saw at least twice a week. To put it another way, you're far less likely to hear about a great job opening from a close friend. You're much more likely to learn about it from a distant colleague. "It is remarkable," Granovetter marveled, "that people receive crucial information from individuals whose very existence they have forgotten."

Why would this be? Surely your close friends are the ones looking out for you, eager to help you find a good job? Sure, but as Granovetter pointed out, your friends have an informational deficit. They're too similar. This is the principle of homophily: Socially, we tend to be close friends with people who mirror us demographically, culturally, intellectually, politically, and professionally. This makes it easy to bond, but it also means that we drink from the same informational pool. Any jobs my close friends have heard about, I've heard about, too.

Weak ties are different. These people are, as Granovetter pointed out, further afield, so they're soaking in information we don't have and moving among people we don't know at all. That's why they're the people most likely to broker juicy job offers. "Acquaintances, as compared to close friends, are more prone to move in different circles than one's self," Granovetter argued. The ties are weak, but they are rich conduits for information.

Of course, back in Granovetter's 1970s, pretty much the *only* time you'd talk to weak ties was when you were on a desperate hunt for fresh information—such as during a job search. Otherwise, it was a hassle to keep in touch with them. The only people who trafficked in tons of contacts back in those days were what Malcolm Gladwell called connectors, those rare, deeply social types who knew far more people than the average. Those people were powerful (certainly at work, and often in everyday life, too) precisely because of their octopuslike connections to so many weak ties. They brokered information that the rest of us were simply too lazy, busy, or constitutionally unsuited to leverage. We had limits on our sociality.

Today's ambient tools dissolve those limits. They make it far easier for us to keep tabs on weak ties and to make more of them. This phenomenon transforms everyday people into superconnectors, in everyday lightweight contact with far more people than before. The explosion of weak ties is how I and my wife found our house. It was how an acquaintance was able to notice—at the corner of her attention—what had happened to Mona Eltahawy and how Suarez's colleagues are able to find answers so quickly at work.

As we've seen, weak links are particularly useful for solving problems, because they harness the power of multiples. You may not know the answer to a question, and your friends may not—but if they repeat your question, odds are *their* friends will, because once you've moved out to the third ring on your social network you're dealing with scores or even hundreds of people.

This creates some spectacular feats of collective answering. Peter Diamandis, the head of the X Prize Foundation, was wondering about the viability of mining asteroids—which led him to speculate on the total volume of gold ever mined on Earth. He did a back-of-the-envelope guesstimate, then posted it as a status update: "Total gold ever mined on Earth is 161,000 tons. Equal to ~20 meters cubed . . . pls check my math!!" In a few minutes three of his followers had verified his estimate, and others had produced estimates for platinum, rhodium, and palladium. ("Ask and you shall receive," Diamandis joked.) You see this behavior everywhere online—late-night parents wondering about their child's fever symptoms, gamers wondering how to beat a boss, tourists looking for a good bar, PC users debugging a crash. Indeed, as a mechanism for finding knowledge, weak-link networks occupy a cognitive role usefully different from, say, search engines. While Google is useful at quickly answering a specific factual question, networks of people are better at fuzzy, "any-idea-how-to-deal-with-*this?*" dilemmas that occupy everyday life. They harness wisdom and judgment, not just pattern matching of facts.

Mind you, acquiring a network that feeds you surprising and valuable knowledge doesn't happen on its own. Like most of our new digital tools, crafting a good set of weak links takes work. If we *don't* engage in that sort of work, it has repercussions. It's easier to lean into homophily, connecting online to people who are demographically similar: the same age, class, ethnicity and race, even the same profession. Homophily is deeply embedded in our psychology, and as Eli Pariser adroitly points out in *The Filter Bubble*, digital tools can make homophily worse, narrowing our worldview. For example, Facebook's news feed analyzes which contacts you most pay attention to and highlights their updates in your "top stories" feed, so you're liable to hear more and more often from the same small set of people. (Worse, as I've discovered, it seems to drop from view the people whom you almost never check in on—which means your weakest ties gradually vanish from sight.) As Pariser suggests, we can fight homophily with self-awareness—noticing our own built-in biases, cultivating contacts that broaden our world, and using tools that are less abstruse and covert than Facebook's hidden algorithms.

If you escape homophily, there's another danger to ambient awareness: It can become simply too interesting and engaging. A feed full of people broadcasting clever thoughts and intriguing things to read is, like those seventeenth-century coffee shops, a scene so alluring it's impossible to tear yourself away. Like many others, I've blown hours doing nothing of value (to my bank account, anyway) while careening from one serendipitous encounter to another. Others have complained that ambient awareness stokes their FOMO—"fear of missing out," the persistent dread that there's some hashtagged "happening" they're missing out on *right this instant*, a sort of hipster recency paranoia on overdrive.

Still, I'm not especially worried about the much-ballyhooed notion of social media addiction, the idea that one might neglect one's "real" life and social contacts in favor of the screen. True cases of Internet addiction are rare, and the panic about social isolation is likely completely backward. As Lee Rainie and Barry Wellman document in their book *Networked*, people who are heavily socially active online tend to be also heavily socially active offline; they're just, well, *social people*. But while the language of addiction is overwrought, it's true that ambient media can be deeply seductive. It creates the gamelike cycle of what psychologists call intermittent reinforcement. Not every status update is interesting, and the really fascinating ones occur only now and then; this tends to drive us mad with curiosity, constantly checking feeds in hopes that they'll dispense a burst of novelty.

The trick here, again, is mindfulness. We need to notice when our dallying in the ambient world is taking us away from other things we ought to be doing. To maintain cognitive diversity, we have to step away from the screen to immerse ourselves in the slower-paced pleasures of nature, books, and "offline" conversation and art. After all the upside of an always-on, teeming world of information is that it's always on; ignore it for the weekend and it'll still be around when you get back. Stowe Boyd, a pioneer in social media, once compared ambient signals to a stream of water. You go to a stream to take a sip—not to try to inhale the entire thing. "You take a drink, and you walk away until you're thirsty again," he told me.

There's another counterintuitive rule here, too. When it comes to ambient contact, smaller can sometimes be better. As with public thinking, sometimes connecting with fewer people is more valuable than connecting with more.

Smaller, of course, doesn't describe the way the most high-profile "power users" behave. Early adopters of ambient tools were in a frantic race to gather as many followers as possible, as a way of building their personal brand and maximizing influence. If you adopt the corporatist logic of treating your personality as a commercial product, what's good for General Motors is good for you. You want as many customers for your utterances as possible.

But people who use ambient signals as a way to *think*, not merely to self-promote, have discovered something different. When an online group grows above a certain size, signals get drowned in noise, and the exchange of ideas grinds to a halt.

Consider the case of Maureen Evans. Evans was a grad student and poet living in Belfast who loved to cook. In the early days of Twitter, back in 2007, she started a witty project to post entire recipes inside individual tweets, using the handle @cookbook. This required feats of compression that read like postmodern poetry; her recipe for lemon lentil soup read "mince onion&celery&carrot& garlic; cvr@low7m+3T oil. Simmer40m+4c broth/c puylentil/ thyme&bay&lemonzest. Puree+lemonjuice." Soon she had three thousand followers, and it felt like a town where clusters of the regulars knew each other, and she could overhear them talking. They'd chat about recipes and argue about ingredients, and some would direct-message her to carry on one-on-one conversations. For the next two years, Evans's feed slowly became more popular, rising to six thousand followers.

Then something happened: Evans became *too* popular. *The New York Times* wrote a story about her @cookbook account. ("A dish, a meal, a trip to deliciousness magically packed into the tiniest carry-on bag," gushed Lawrence Downes.) Because of the publicity, within a month her Twitter count more than tripled, shooting up to nineteen thousand—and then kept on growing, eventually cracking over two hundred thousand people. For Evans, the exposure was fun (and useful; she got a book deal to write *Eat Tweet*, a collection of her recipes). But it came at a cognitive cost, because the conversation in her feed died. People stopped crosstalking; when she posted a recipe or a comment, the smart people stopped directly replying to her. "It became dead silence," she tells me.

Evans ran into the same problem newspapers hit with their comment threads: Socializing doesn't scale. Conversation doesn't scale. "I think when you see yourself as a member of a huge group, you're not as confident in your own voice and the efficacy of your contribution," Evans says. "You feel you can't possibly be the person who's going to make the useful contribution in the conversation." I've heard related complaints from people who enjoy engaging in at-reply discussions with their followers, but then suddenly become Twitter famous. The proprioception of the group breaks down. If a few dozen, a few score, or even a few hundred people are at-replying to you, it's possible to respond to many or all of them. But once thousands are doing so, reciprocity gets hard, even impossible.

The lesson is that there's value in obscurity. People who lust after huge follower counts are thinking like traditional broadcasters. But when you're broadcasting, there's no to and fro. You gain reach, but lose intimacy. Ambient awareness, in contrast, is more about conversation and co-presence—and you can't be co-present with a zillion people. Having a million followers might be useful for hawking yourself or your ideas, but it's not always great for thinking.

Indeed, it may not even be that useful for hawking things. In 2010, Meeyoung Cha, then a researcher at the Max Planck Institute for Software Systems, published a paper analyzing over 54 million Twitter accounts, ranging from massive celebrity accounts to ones where people were followed by barely a handful of friends. She wanted to know: Is there any correlation between having tons of followers and being influential? Cha decided to measure how often people's tweets were retweeted, arguing that it was a good proxy for one's overall influence. Being retweeted, after all, meant that people were directly passing on your words to other people—a classic two-step flow of influence, ripples in the pond radiating from your original splash. If having tons of followers was important, then surely those Twitter superstars would always be generating the most retweets.

Yet they weren't. Cha found there was little clear correlation between how many followers you had and the likelihood of being retweeted. Your huge follower count might make you feel important, but, as Cha wrote, it "reveals very little about the influence of a user." Borrowing a term from the software developer Adi Avnit, she called her paper "The Million Follower Fallacy": having a ton of followers doesn't necessarily mean they all care, or even notice, what you say. Or as the technology thinker Anil Dash wryly notes, "they can't count as 'followers' in any meaningful sense."

In contrast, Cha found a set of people who *did* get retweeted disproportionately often. Who were these folks? They weren't necessarily the most popular accounts. But they were experts on a particular subject—people who'd built up a reputation for delivering a regular stream of smart, informative tweets and links on their pet topic. "It's about authenticity and bringing new information into the network," Cha tells me. "Those are the people whose message actually spreads." Quality, delightfully, seems to matter.

"This is no good," Samantha mutters. A recent college graduate, she's in a New York coffee shop called Think, and she's "detagging" herself from photos that friends took at a party and posted on Facebook. They're not particularly racy shots; actually, they look to me like rather sweet pictures of her with her arms around friends. But she and her friends are all holding drinks, and Samantha is starting a job at a conservative, old-fashioned firm—so why risk having these things crop up when someone searches for her name?

Meanwhile, next to her is her friend Andrea, who shows me her Tumblr and Twitter streams. They're mostly devoted to her obsession with interior design, a field in which she hopes to eventually work, but her Twitter stream also includes tons of personal notes mixed in, too—shout-outs to friends, WTF-style commentary on the day's news, cooing over a viral video they all passed around. She doesn't think it's a problem mixing her personal and professional streams. "It's a field where people are creatives," she adds, so they expect idiosyncratic personalities.

Lucas, a recent law school grad, is also on the job market, "so

I'm pretty careful what I put out there," he tells me. I'm less sure: As we flip through his photos on Facebook it feels like a chronicle of ardent binge drinking. Hilariously, I even spot a photo of Lucas at a party of soused people swilling massive mugs of beer—a tableau that has become the archetype of Stuff That Will Get You in Trouble When the HR Guy Googles You. "It's not *that* bad, right?" he laughs. They're set on "private" mode, or at least he thinks they are; suddenly a bit nervous about it, he starts poking around in Facebook's settings to check.

This sort of free-floating anxiety—who's going to see this stuff?—is a defining part of living online in the status-update world. It's particularly acute for young adults, who've spent much of their lives living, to some degree, in public. The anxiety is the flip side of ambient signaling: An utterance that seems innocuous *now* could wind up making you look foolish years later. Sure, a thousand tiny signals build up into a rich picture of your personality and thinking—but if someone rips one of those out of context, it can wreck a reputation. Indeed, innocuous stuff always has. "If you give me six lines written by the hand of the most honest of men, I will find something in them which will hang him," as the seventeenth-century French statesman Cardinal Richelieu is reputed to have said.

"It gets complicated quickly," says danah boyd, a researcher at Microsoft who has done pioneering studies of how online socializing affects people's lives, when I interview her. As she points out, the challenge that comes with broadcasting bits of your life is that you live with an "invisible audience;" you're never quite sure *who's* going to see this stuff, or when. Since anything online can be copied and circulated, you also get what boyd calls "context collapse": sexy talk meant for your partner gets seen by your mother when you accidentally mismanage the settings on your social tools.

In *1984*, Orwell popularized Jeremy Bentham's concept of the panopticon, a technological structure in which powerful forces can watch you but you can't see them. This forces you to assume that you're always being watched, a mental head game that deforms behavior. Today, plenty of similarly powerful forces eagerly observe your online behavior. Firms like Facebook are free only because they carefully collect our ambient signals, the better to sell us to advertisers. (As the joke goes, if you're not paying for a service, it's because you're the product.) Indeed, Facebook CEO Mark Zuckerberg explicitly favors context collapse. He has said he believes there's something shady about having different sides to your personality. "The days of you having a different image for your work friends or co-workers and for the other people you know are probably coming to an end pretty quickly. . . . Having two identities for yourself is an example of a lack of integrity," as he told David Kirkpatrick in *The Facebook Effect.* Of course, I'd argue that history and philosophy show that the opposite is more often true—that it's emotionally and intellectually healthy to have different identities. "We do not show ourselves to our children as to our club companions, to our customers as to the laborers we employ, to our own masters and employers as to our intimate friends," the psychologist William James observed over a century ago. But the corporate logic of data mining and ad delivery sees this sort of healthy, multifaceted identity as a bug, not a feature. Or as Eric Schmidt, the executive chairman of Google, once famously said, "If you have something that you don't want anyone to know, maybe you shouldn't be doing it in the first place," a statement nearly Soviet in its creepiness.

It's no doubt true that corporations and governments love the spectacle of a society outgassing its personal information into the digital ether. (Just ask the insurance companies or hiring committees who can now easily find oodles of info on the private behavior of people they're scrutinizing.) But as boyd and other researchers have found, these problems, while real, are still in the minority. The much more common complications come not from powerful interests surveying your ambient info from above but from those surveying your life beside you: colleagues, ex-lovers, friends, or family. We live less in a panopticon than an "omniopticon," as Nathan Jurgenson and George Ritzer put it, a situation in which we're made uneasy by potential ill uses of our info in the hands of these peers. Our problem isn't just Big Brother—it's all the Little Brothers and Sisters. These are the problems of what New York University professor Terri Senft deftly calls micro-celebrity. Andy Warhol quipped that in the future, everyone would be world famous for fifteen minutes. But micro-celebrity, as Senft notes, follows the rule set down by the blogger and musician Momus: "In the future, everyone will be famous to fifteen people."

Of course, these dilemmas aren't completely new. People have been neurotically managing their reputations for eons. In 1959, Erving Goffman wrote *The Presentation of Self in Everyday Life*, a study of the ways in which people enacted versions of themselves at work, in public, even at home. Goffman likened life to a theatrical performance in which we are each engaged in the craft of "successfully staging a character." Goffman's book depicted a society of 1950s folk frenetically obsessing over how they appear to others: young women dutifully concealing their intelligence from their duller dates, job candidates rejected for not being a "Hollywood type" or having teeth that were too square. Indeed, the good old days of the 1950s were, as historians and authors remind us, replete with a level of omniopticonic surveillance that often out-Facebooked Facebook. But at least in the past, leaving your past behind was easier, such as if you fled your small town and moved to the big, anonymous city. Plus, fragile human memory meant scandalous details faded organically over time. Today, behavior and utterances remain enshrined, ready to pop back up with perfect fidelity years later.

The result is a palpable tension across our ambient lives. Broadcasting our thoughts and experiences is at once valuable and risky. But since the benefits are immediate and powerful, and the risks merely uncertain, we tend to stick with it. Is there any way to reduce the risks?

Some young people I talk to, those who've grown up with ambient signaling, predict that cultural change will eventually defang some of this problem. If everyone posts pictures of their youthful revels online, in twenty years the "running for president" problem may evaporate. You can't derail someone's bid for political office by unearthing their old beer-bong picture if *everyone* has posted beer-bong pictures. "I just think this isn't going to be as big an issue by the time everyone my age is in their forties," as a nineteen-year-old student told me (though ironically she asked me to not use her name in print). Her attitude might seem naive, except that similarly seismic cultural shifts have occurred in the past. Fifty years ago, divorce was still sufficiently shameful that it wasn't talked about in polite conversation, and being divorced was a barrier to seeking public office. Today, the stigma against divorce has so evaporated that the U.S. Republican Party, an entity expressly concerned with the erosion of the American family, routinely contemplates candidates who've been divorced, sometimes more than once. (John McCain had one ex-wife; Newt Gingrich had two.) The same is beginning to happen with marijuana, a drug Barack Obama publicly discussed using not long before announcing his candidacy for president. (When asked if he'd inhaled, he said, "That was the point.") There's a counterargument here, though: Society may retire old forms of social censure, but it's very good at coming up with *new* ones, particularly for young women. Even if the beer-bong taboo recedes, society will find behavior to slam down. There will always be private behavior that we'll want to keep private.

We'll need to become more adroit at using ambient tools—and remain critical users. Our record thus far is mixed. People have a long track record of handing over power to the Facebooks of the world in exchange for convenience. On the optimistic side, some of the most immersed users of this stuff, teenagers, have pioneered skillful ways of hacking their digital forums. "It's completely untrue that 'teens don't care about privacy,'" boyd tells me. "They're actually very careful [about] what they put online and what they don't." Indeed, boyd's research documents young people who've cleverly found ways to wrest back some control of their digital lives. One student, seventeen-year-old Shamika, "white walls" her Facebook account by deleting updates more than a day old. This lets her use the account to broadcast ambient signals to friends but mostly keep them from being archived. Others students have devised an even more clever strategy: They resign their Facebook accounts when they want to vanish from sight. When you resign, your account becomes invisible. To the outside world, it's gone. But since Facebook hopes you'll come back, it doesn't actually delete your information. It's still sitting there, waiting for you to start it up again. So these teenagers run their Facebook accounts like salons or bars, with opening hours. They reinitiate their account for an hour or so, during which time they can post photos or updates and chat with their friends. When they're done, they resign again, and their online presence goes dark, preventing anyone from rummaging through their postings.

If we're very lucky, and this certainly isn't a given, public demand will propel a wider range of ways to interact ambiently—some more private, some less. Facebook seems dominant now, but that could change quickly if someone else comes up with a service that better suits people's desire to control their online presence. Many alternate spaces already thrive. Indeed, some of the most interesting conversations I encounter online happen in smaller-scale, less-populated discussion forums devoted to hobbies. Because small forums cost almost nothing to run, hobbyists don't need to succumb to the privacy-eroding logic of monetization. And in smaller, more intimate groups, participants worry less about context collapse. Other forums—like the infamous 4chan or the mothers' board YouBeMom—offer radical anonymity as a way to encourage people to speak freely, and the conversations there are more freewheeling, for good and ill, than just about anywhere online.

The culture of being always on—available to respond to any social ping, and feeling compelled to do so—may also fade. The anthropologist Genevieve Bell thinks our early infatuation with incessant online contact is already easing. She's found that some young people shut their phones down—for mental peace, privacy, or even just because it marks them as cooler than the continually connected. "The moment we're in, where everyone has three devices out and is conducting five conversations with people who aren't there— I'm not sure we're going to be doing that five years out," as Bell says to me. Mobile phone calls have already gone through that curve. For the first decade of their mainstream adoption, people answered them literally every time they rang—at dinner, at a funeral, while having sex—because they were so intoxicated by the feeling of instant contact. But after a while it became clear that this behavior was annoying, and it faded.

Technologically, there are plenty of ways to design tools so they diminish the omniopticon. There's Viktor Mayer-Schönberger's proposal for what we could call artificial forgetting—designing sharing tools so they automatically delete photos or posts after a period of time. As he argues, this would let us enjoy the benefits of ambient intimacy without accumulating archives of our behavior. As proof this can work, consider the example of Drop.io, a service (now sadly discontinued) that I used for sharing files. When you uploaded a file, Drop.io would ask you when you wanted it deleted: a day from now? A month? A year? Crucially, deletion was the *default*, the inverse of most save-everything Web services. If you wanted something left permanently online you had to specifically set it that way. This behavioral hack worked. When I interviewed then-CEO Sam Lessin, fully two thirds of the material people had uploaded to Drop.io in the previous year and a half was gone. More recently, social media tools like Snapchat or Poke have emerged, with the aim to make individual bits of sharing—a photo, a note—vanish after a single view. Obviously, artificial forgetting can't stop malicious people from cutting and pasting personal info elsewhere. But it minimizes the windows in which this happens.

There's also the possibility that new laws and regulations could help us minimize unwanted digital traces, at least in countries that respect the rule of law. Harvard law professor Jonathan Zittrain has suggested a form of "reputation bankruptcy"—every decade you're allowed to request the deletion of your online utterances from major services. A group of his students proposed a similar regulation aimed specifically to protect minors. And the European Commission is considering a "right to be forgotten," similar to Zittrain's concept. These are imperfect solutions that coexist uncomfortably with free speech laws in liberal democracies, but they're worthy of debate. And liberal democracies have, after all, navigated similar legal shores before. When tools and institutions emerged to track Americans' personal data—like the Social Security number or credit bureaus—we passed laws to limit who could see our info and how long it was retained. Artificial forgetting can work if we want it to.

It's worth thinking about these trade-offs and countermeasures, because ambient awareness is likely to continue growing—in new and increasingly strange ways.

Geography has already become a frequent status update, as millions of people worldwide use GPS to broadcast their locations, providing a rolling map for friends ("I'm at Bar X!"). Cultural experiences are beginning to flash their own ambient signals, as with music fans whose music players automatically report online what they're listening to; as this becomes geolocated we'll be able to see what types of music and moods are being generated all around us. Some technologists have even experimented with caps that can track and broadcast crude measures of brain-wave activity, so a professor could "see" whether his students were confused by his lecture, or you could read the mood of a bar before going in.

We'll likely also see more *things* broadcasting status updates. Right now, most ambient signals are actively composed. But as sensors become cheaper, we'll discover that more of our tools and possessions can, if we want, tell the world how we're using them. Think of those runners who use Internet-connected pedometers to autoreport how much they've run, or dieters who use connected bathroom scales to tweet their weight publicly. Why on earth would they do something so revealing? To harness the audience effect: When we know someone's watching, we're more likely to stick to our vow to lose weight or run. When we see what other people are doing, they inspire us. But frankly, sharing can also be pleasant. It adds social glue to our acts of grinding discipline.

Even old-fashioned forms of contemplation, like book reading, are likely to develop ambient signals. Readers are already streaming "highlights" of what they're reading onto services like Goodreads, Findings, or Amazon's Kindle. As with all forms of floating awareness, this is simultaneously unsettling and thrilling. Unsettling, because the privacy of reading—protected by confidential library borrowing records—is a civic virtue; it allows us to explore controversial or unpopular ideas without the scrutiny of authorities or our peers. But thrilling, too, because when you broadcast your book reading voluntarily, it creates moments of fascinating serendipity. Strangers strike up online conversations after noticing my public book notes; friends would discover what I'd been reading and talk about it when we met. After broadcasting my reading for several months, I began wishing for X-ray vision into the reading around me. I'd walk onto a subway car or into a café and wish I had an app that would show me whatever page my seatmate was reading in her e-book, while I'd show her mine. Or I'd wish I could locate someone nearby reading the same book, so we could chat. It's like an update on the age-old habit of noticing the dust jackets on books others are reading: a momentary glimpse into the life of another mind.

The Connected Society_

In China, they're known as the "post-'90s," and they don't have a very good reputation.

They're youth who were born in the 1990s, and in the eyes of Chinese liberals and intellectuals, they're regarded as quintessential slackers. They grew up in a world shaped by Deng Xiaoping's go-go capitalism, in which the Communist Party had erased the 1989 Tiananmen Square revolt from history books and the local Internet. Politically ignorant and coddled by their parents—they are all "only children," courtesy of China's one-child-per-family policy—these post-'90s kids were regarded as feckless and materialist, interested only in video games, wasting time online, and fashion (such as their "exploded head" hairstyles). If you were a liberal in China who longed for democracy, you looked at this feckless generation and sighed: the death of politics.

Until the summer of 2012, when that myth came crashing down—as the post-'90s youth of Shifang staged one of the most successful environmental protests ever.

Shifang is a city in southwestern China. In late June 2012, local party officials announced they would begin building a $1.6 billion plant to process molybdenum-copper alloy. The plan was slated to create thousands of construction jobs in an area that was in distress, because of both the country's economic slowdown and an earthquake that hit four years earlier. But many locals worried about toxic side effects. Copper plants often produce slag filled with noxious chemicals, including arsenic; many such areas have seen cancer rates soar. Chinese citizens are painfully aware of how rampant development is ruining the countryside. In some industrial cities the air is so foul that people wear surgical masks outside and rivers run in different colors each day of the week. Still, the plant looked inevitable; a kickoff ceremony was set for June 29. Locals might grumble, but it was rare for party officials to back down—particularly since those officials usually grew wealthy from kick-backs.

This time, though, Shifang's students struck back.

In the days leading up to June 29, the students began networking online. They met on QQ, a popular instant-messaging service, as well as discussion boards hosted by Baidu, the country's main search engine. Meanwhile, they began seeding Sina Weibo—China's enormous social network, roughly comparable to Twitter—with doubts about the copper plant. "Overdose of molybdenum may cause gout, arthritis, malformation, and kidney problems," one user noted. Another argued, "Without doubt, Shifang will become the biggest cancer town in years," and pleaded for others to rise up: "Who can help people like me who don't want their hometown to become hell?"

The students decided they would hold a protest in Shifang's streets on July 1, since, in a nice bit of irony, this was the birthday of the Chinese Communist Party. They'd meet in front of the Shifang municipal government offices. As the online criticism about their plant grew, the students carefully spread the word offline, too—handing out flyers and printing T-shirts to broadcast their cause.

When the day arrived, the protest started small, with the early crowd of students in the dozens. Toting colorful umbrellas against the rain, they sported banners and T-shirts saying "Protect the environment at Shifang. Give my beautiful hometown back," and "No

More Pollution! We'd Like A New Shifang." But while the protest started small, it soon ballooned. Students and interested locals posted pictures on Weibo, where they were quickly recirculated. Meanwhile, the students began mass texting other townspeople: "People of Shifang, be united and together protect our home!" they wrote. "Many students and residents gathered in front of the government building, many of which skipped classes or acted against their teachers' persuasions. . . . It seems that the influence of these students are not strong enough! I ask all of you and your relatives to drive out your cars, block the traffic and make it big to pressure the government. This is a choice out of no choices!"

By the end of the first day, the crowd had grown to five thousand. The next day it was even larger—some reports had it doubling—and it wasn't just students anymore, but citizens from across Shifang, including shopkeepers and businesspeople. Via Weibo, the entire country was learning of the protest; "Shifang" was the top searched-for term.

Local officials panicked. The protest had come out of nowhere, catching them off guard. They ordered riot police to break it up, firing tear gas canisters and striking demonstrators.

This action backfired badly. Not only did it fail to break up the protest, but students and Shifang citizens instantly uploaded pictures and video of the crackdown, flooding the country with ghastly images. One showed a man with a huge open gash on his left shoulder; in another, a woman's mouth was covered in blood; a third showed an infant who'd gotten blood, probably his parents', smeared across his forehead. Then came stirring images of bravery: a young woman in a blue dress, arms outstretched, confronting an officer; another woman kneeling before a cluster of riot police. "There are too many 'first times' today," posted one young female protester: "Except during menstruation, it's my first time to bleed so profusely. First time to see my own bone. . . . We simply hope that our hometown is free from pollution. That's all. Is that too much to ask?!"

Normally, the government carefully blocks this sort of dissent, ordering Chinese social networks to sniff out illicit keywords and delete postings. And indeed, some Shifang postings were deleted. But many more got through, possibly because the sheer volume was too overwhelming for the autocensoring software, and images can't easily be autodetected. It's also possible the central government worried that fully banning the chatter would cause yet more controversy. Either way, a study by the China Media Project found that in the four days there were a stunning 5.25 million postings about the protest on Weibo. Most of these were rebroadcasts of fifteen thousand original bits of on-the-scene reporting by the protesters themselves. In a fruitless local attempt to stop the flood, Shifang's police issued a warning: "The use of the Internet, cell phone SMS messages or other means to organize or instigate illegal assemblies, protests or demonstrations of any kind is strictly prohibited."

It was too late. On the third day, local officials admitted defeat. They publicly announced they were canceling the copper plant.

It's hard to know why the officials decided to back down; the Communist Party doesn't discuss its meetings openly. But observers suspect the sheer weight of public attention was too intense, even for a party that has no opposition and routinely throws dissidents in jail. Plus, while democracy advocates can be marginalized, environmental issues are harder to brush off. "The information and pictures shared through Weibo aroused national attention," Ma Jun, founder of the Institute of Public and Environmental Affairs, told *Businessweek*. Or as one Weibo user noted: "Before, no one knew what the real situation was, or the people's discontent couldn't make it onto the nightly news. But now, it's different. With the Internet, the government's job just gets harder and harder."

The students were exultant. "It is the 4th of July—236 years ago

America achieved independence, and 236 years later, the Shifang people are fighting for their own rights and confronting the government," a microblogger wrote. A female student from Sichuan Engineering Technical College posted, "Joining this protest is the most meaningful thing I've done in my 19 years of life."

The idea that technology can help liberate the oppressed has long been a seductive one. As Elizabeth Eisenstein argued in *The Printing Press as an Agent of Change*, the printing press allowed economies of scale that spread liberating ideas far more broadly than was before possible. (Indeed, Martin Luther's ninety-five theses were so rapidly reprinted—across various countries—that Luther himself was stunned. "It is a mystery to me," he admitted to the pope, "how my theses, more so than my other writings, indeed, those of other professors, were spread to so many places.") Anticommunist politicians and intellectuals in the West have long credited samizdat publishing for helping erode the Soviet Union.

But the Internet and modern digital tools have been particularly romanticized. Our ability to communicate instantly across vast distances, to speak to the world and to each other, seems uniquely *freedomish*. In the West, in particular, we tend to regard speech as both a proxy for emancipation and its inevitable catalyst; and if the Internet has done anything, it's produced a global flood of speech. Prophecies of liberation have come fast and furious. "The Goliath of totalitarian control will be rapidly brought down by the David of the microchip," Ronald Reagan intoned in 1989. And as Hillary Clinton more recently proclaimed, the United States was willing to "bet that an open Internet will lead to stronger, more prosperous countries."

Alas, this isn't true. Or at very least, it isn't *simply* true. Communications tools may be a necessary condition for broad-based social change, but they aren't a sufficient condition. If the advance of human rights was simply a matter of sprinkling around microchips and bandwidth, several of the world's most despotic regimes ought to look like Canada. In reality, results have been erratic. In some countries where use of the Internet and mobile phones has boomed, there have been significant shifts toward freedom and rights, notably in Arab Spring states such as Egypt and Tunisia. In others, such as Azerbaijan and Belarus, the growth of the Internet has been accompanied by brutal crackdowns. And some countries, such as China, tread a middle path. The Communist Party filters activity online, running a Great Firewall that blocks citizens from seeing foreign news the party deems seditious, while threatening Chinese Internet executives if they don't self-police their users. But because China also needs the Internet and mobile phones for economic growth, the party has tolerated and even encouraged their spread. (Indeed, in some areas of China you can get broadband wireless data more reliably than in my gentrified corner of Brooklyn.) These digital tools haven't toppled the party, but the sea of online chatter has accompanied a striking increase in traditional offline protest. Twenty years ago, the number of demonstrations in China was 8,709 per year. By 2012 it had ballooned to about 90,000. While these protests are driven by primal, real-world needs—cleaner air, better pay, less corruption—the increasingly networked nature of Chinese society is likely a big part of the shift, as the protests by the students of Shifang show.

To understand how technology affects social change, you have to look at how it affects the way we think, learn, and cooperate with others, and how local cultures come into play. All the enhancements of our cognition—bigger memory, public thinking, new literacies, and ambient awareness—play critical roles in how political change unfolds and how it's thwarted.

Actually, one of the best ways to grasp how political change happens is to study how it *doesn't*.

Think of a situation where there's a major social injustice staring everyone in the face, yet nobody protests. Take the example of racial segregation in the United States in the late 1960s. On paper and legally, the country by then had made most forms of segregation illegal. The Fair Housing Act of 1968, for example, made it illegal for landlords to discriminate on the basis of race. Polls of American whites showed that they generally agreed with these changes—indeed, clear majorities believed official segregation was wrong. Yet in everyday life, segregation was common: White employers routinely wouldn't hire blacks, and white neighbors often opposed blacks moving into their neighborhoods. The private views of whites—that segregation was bad—wasn't affecting everyday behavior. Why?

Back then, sociologist Hubert O'Gorman got interested in this puzzle, and he did some clever research to untangle it. First, he examined a national survey that asked white Americans across the country to state whether or not they supported segregation. Sure enough, the large majority didn't. Only a small minority of whites—18 percent of the total—still believed in segregation.

Then O'Gorman considered a second question: How many whites believed *most other whites* supported segregation? In other words, maybe you weren't racist . . . but did you think most of your neighbors were? Sure enough, many did. Nearly half of all whites—47 percent—believed that the majority of white people were segregationists. In essence, lots of American whites held an amazingly inaccurate mental picture of what other whites believed.

This, O'Gorman reasoned, helped explain why racist behavior

and discrimination were still so widely tolerated. Because most whites inaccurately believed they were surrounded by tons of other white racists, they were less willing to speak out and oppose racism. This was particularly true in areas like housing. When O'Gorman looked into whether whites supported segregated neighborhoods, he discovered that only a very small minority of hard-core racists favored it—and an equally small minority heatedly opposed it (the social-justice folks, as it were). But the large swath of middle-of-the-roaders? They weren't particularly racist, but because they incorrectly believed that *most white Americans* supported segregated housing, they supported it, too. (Now, there's an obvious objection to O'Gorman's analysis here. Isn't it possible that lots of people were simply lying about their racist beliefs to the survey takers, trying to appear more tolerant? Sure, but O'Gorman didn't think the data supported that. The people most likely to lie—those living in liberal northern areas—also had the lowest rates of false beliefs.)

This same dynamic, O'Gorman argued, helped explain why blacks couldn't get hired: Even when white employers did not think of themselves as racist, they believed their customers were—so they went along with the crowd, or at least the crowd they imagined existed. O'Gorman quoted the economist Gunnar Myrdal's interviews with white employers: "I have been told time and time again that they have nothing against employing Negroes, and I believe they are telling the truth. What holds them back are the considerations they have to take about the attitudes of customers and co-workers." The same dynamic even appeared to govern playdates of children. In a 1969 survey, fully 76 percent of whites polled in Detroit said a white mother should let her daughter bring a black friend home from school. But only a third believed other whites would agree, so few such playdates took place.

Sociologists have a name for this problem: pluralistic ignorance. It occurs whenever a group of people underestimate how much others around them share their attitudes and beliefs. It's a huge impediment to social change. It happens for psychologically understandable reasons, of course. The social norms of the past can loom large in our memory, even as they're actually dying off person by person.

Pluralistic ignorance isn't just a problem with racism. It affects nearly every domain of human activity. For example, a 1980 survey found that even though a majority of people wanted more laws to limit pollution and nuclear power, they believed others were less supportive. In corporations, board members might all privately notice the lousy performance of the CEO, but none speak out because they don't think the others agree. And university campuses are hotbeds of pluralistic ignorance, particularly with regard to sex. Research shows both both male and female students overestimate how willing other students are to have sex.

Pluralistic ignorance is an information problem, as Andrew K. Woods, a legal scholar who introduced me to the concept, points out. It happens because we don't know what's going on in other people's minds. Whenever we're faced with a socially dicey, delicate subject—Do other people notice that this company is in trouble? How much sex are other students having?—we're too squeamish to talk openly. Without correct information, we get it wrong.

But the converse is also true. It turns out that you can fight pluralistic ignorance by actively improving the flow of information—and letting people know the previously invisible views and thoughts of others.

This is how U.S. campus officials started fighting binge drinking among students in the 1990s. Studies found that binge drinking was a product of pluralistic ignorance: Students assumed their peers drank more than they actually did. In Montana, for example, men thought the average number of drinks a male consumed in a single occasion was seven; in actuality, it was three. Women estimated five drinks for women, when it was really two. So some students would ramp their drinking upward to try to match this mistaken norm, with dangerous results. To fight the problem, officials at Northern Illinois University set up an awareness campaign. They plastered the campus and its widely read newspaper with notices showing the real averages, not the mythical ones, and offered small rewards for students who could produce the right numbers. Sure enough, students adjusted their behavior. Over the next decade, the proportion of students who engaged in high-risk drinking plunged from 43 percent—above the national average of 40 percent—down to 25 percent.

To make social change begin to snowball, we need to make our thoughts visible. When members of society think in public and keep in ambient contact with one another, it creates a new environment—where we're increasingly aware of what changes might be possible.

You can see this dynamic in some of the more dramatic events of the Arab Spring, like the Egyptian uprising.

It wasn't a mystery why Egyptians wanted to revolt. They had plenty of reasons. During Hosni Mubarak's thirty-year presidency, the quality of life had plummeted as the price of food rose, minimum wages went down, and corruption ran rampant, with Mubarak's regime privatizing government agencies and pocketing kickbacks. Those who challenged the regime were dealt with ruthlessly. But behind the scenes, pro-democracy activists and workers patiently toiled to foment dissent. Many groups formed, such as the Egyptian Movement for Change—dubbed Kefaya, meaning "enough"—that arose in 2004, explicitly demanding that Mubarak step down. Incensed by low wages and privatization, workers staged strikes that brought up to twenty-seven thousand people into the streets in 2006. Inspired by this worker discontent, young tech-savvy activists like Ahmed Salah, Ahmed Maher, and Walid Rachid began using online tools to organize protests, in what was known as the April 6 Youth Movement. Meanwhile, the journalist Wael Abbas collected videos of police beatings and torture and posted them online. These activists were plugged into the larger wired world of young Arab dissidents, from Tunisia to Bahrain. Talking on blogs and meeting at conferences, they swapped ideas on how to get their message out while staying safe. They were willing to put their bodies on the line for the cause, getting arrested and beaten for their efforts.

Still, all this activism never ignited the Egyptian mainstream. In fact, most people didn't know it was going on, because Mubarak controlled the state press. "Most of their marches and protests attracted only a few hundred, which made it easy for the police to crack down on them," as the academics Sahar Khamis and Katherine Vaughn noted. As a result, many everyday Egyptians became convinced change was impossible and dangerous to discuss, as Wael Ghonim, an Egyptian who played a role in the uprising (and worked for Google at the time), has written: "Fear was embodied in local proverbs, such as 'Walk quietly by the wall (where you cannot be noticed),' 'Mind your own business and focus on your livelihood,' and 'Whosoever is afraid stays unharmed.'"

Through violence and fear, despots actively *create* pluralistic ignorance. They work hard to engender a lack of trust and knowledge between fellow citizens. This helps derail protest before it can begin: Nobody wants to show up for a demonstration if only a few other people show up, because you'll be jailed—and worse, nobody will ever hear about it. A huge protest is a different story; if you can get hundreds of thousands of people and massive public awareness, you have more power and safety. But for a public uprising to become huge, people have to believe many, many other fellow citizens will conquer their fear and show up. It's a collective action problem. How do you get the snowball rolling?

In Egypt, an event in June 2010 helped break that collective action problem: A young man named Khaled Said was viciously beaten and killed by state police. Egyptian political blogs were alight with graphic photos of Said's brutalized body; Ghonim, the Google executive, was living in Dubai when he saw them. Incensed, he set up a commemoration page on Facebook called "We Are All Khaled Said." A year earlier, Facebook had released an Arabic version whose membership had grown to 3.5 million users, clustered mostly in the major cities. Ghonim's first post read: "Today they killed Khaled. If I don't act for his sake, tomorrow they will kill me." Within two minutes three hundred people had joined the page, and in an hour, three thousand. By the end of the day, it rose to thirty-six thousand people, who'd posted an astonishing eighteen hundred comments, decrying Said's death and venting anger at the government. Ghonim began posting nearly twenty-four hours a day, seeding the page with reports of corruption and torture, as well as rebutting the regime's claims that Said was a drug user. Within four days, one hundred thousand people had joined the page.

What Ghonim noticed was that the crowd wasn't just reading and "liking" his posts. They were talking among themselves, sharing stories of their discontent. This was new behavior. There had been online forums in Egypt, but never one that reached so widely beyond the activist community. In Egypt back then, a very popular Facebook page would have only about one hundred comments, as Rasha Abdulla, an Egyptian professor of communications, said at a New York conference months later. This forum, in contrast, was the first to gain real traction. It was also personal: because Facebook required real names, the posters could see that average Egyptians, friends and colleagues, were equally up in arms.

Ghonim and the other administrators of the page began nudging the community toward collective action. He asked them to post pictures of themselves holding up signs that said "Kullena Khaled Said" ("We Are All Khaled Said"), to give a visceral sense of who they were. Then one member suggested they assemble in a silent protest along the sea in Alexandria. Thousands of page members showed up there and in other cities. Ghonim used Facebook's polling tools to see if members wanted to stage more vigils; they voted yes.

Soon, they had another powerful spur to action. In Tunisia in December 2010, a street vendor died after setting himself on fire to protest police corruption; it sparked antigovernment protests that, in mid-January 2011, drove the president, Zine El Abidine Ben Ali, to step down after twenty-three years in power. The spectacle electrified Egyptians. To capture the energy, Ghonim, working with other activists, suggested a massive protest for January 25. Quickly, twenty-seven thousand members of the page RSVP'd to say they would come out; discussion exploded to fifteen thousand comments a day, and the page ballooned to five hundred thousand members, making it the largest online protest group the country had ever seen.

In other words, the collective action problem was dissolving. The members of Ghonim's enormous group showed up because they knew they weren't alone. And other activist groups were amassing similar gatherings. January 25 became history, the beginning of a mass uprising that drove Mubarak from power only two and a half weeks later.

"People who would only post comments in cyberspace became willing to stand in public; then those protesters, among many others, made the great leap to become marchers and chanters, and grew into a critical mass that toppled a brutal and tyrannical regime," as Ghonim later wrote.

Egypt was a perfect example of how deeply online communications have become embedded in social change. It is silly to call the uprising, as some pundits did, a "Facebook revolution." Revolutions happen when people put their bodies on the line, not merely fingers on a keyboard. (Indeed, about 850 civilians died in the conflict.) And the power of offline, real-world social connections was crucial. For example, an element in the revolution's success was the participation of Egypt's soccer fans, the Ultras—because after years of rioting during games, the Ultras were expert in fighting police and staging public spectacles.

Yet ordinary Egyptians wouldn't have been prepared to show up in the first place until they'd burned off their pluralistic ignorance by communicatng with one another. "Once you know other people agree with you, it just cascades from there," as Zeynep Tufekci, a Turkish-American academic who carefully studied the Egyptian uprising (and who appeared in the previous chapter, aiding Mona Eltahawy), tells me. Two weeks after Mubarak stepped down, Tufekci and her research team polled 1,050 Egyptians, some of whom had shown up on the first day of protest. She wanted to know how precisely they'd first heard about the event. Tufekci found that person-to-person contact was key—offline and online. Forty-eight percent had heard about it directly from someone they knew, which is what you'd expect. But the next biggest chunk, 28 percent, first heard about it from Facebook. Almost none heard about it from traditional broadcast media.

In a sense, activism has always been about fighting pluralistic ignorance. A key job of the public organizer—in a community, at a workplace—is to catalyze existing discontent, to persuade people that others share their visions and desires. Nor is the idea of a public talking to itself anything new. "Speaking, writing, and thinking involve us—actively and immediately—in a public," as social theorist Michael Warner notes. What is different is the scale and speed at which publics can form and reach awareness.

You can see these effects of scale and speed in the earliest days of electronic communication. When the electric telegraph emerged in the nineteenth century, one of the first things businesspeople sent from town to town were reports of local weather; for corn speculators, this information was crucial. But these weather reports had a curious secondary effect. As people became conscious of weather around the country, they began to see it as a system with patterns, as the author James Gleick writes in *The Information*. "The telegraph enabled people to think of weather as a widespread and interconnected affair, rather than an assortment of local surprises," he adds. This is precisely what happens now with politics: Citizens can see patterns of common grievance emerge locally, nationally, and internationally.

One of the dangers of online conversation is if it *remains* conversation, never turning into action. Complaining is easy—much easier than getting out of your chair. Many critics have worried about the rise of so-called slacktivism, a generation of people who think clicking "like" on a Facebook page is enough to foment change. Dissent becomes a social pose.

Indeed, Egyptian activists fretted about this. A week before the January 25 rally, Asmaa Mahfouz—one of the young leaders of the April 6 movement—posted a stern YouTube warning. "Sitting at home and just following us on news or Facebook leads to our humiliation . . . If you stay home, you deserve what will happen to you," she said sternly. Commenters on the "We Are All Khaled Said" page echoed her fears. Even they didn't think anyone would step away from the screen. "No one will do anything and you'll see," one fretted. "All we do is post on Facebook. We are the Facebook generation. Period."

These fears weren't borne out. It's true that people can be lazy and that motivating people to do real-world work—attending protests, donating money, making phone calls—is hard. But it was always this way. Long before the Internet, when I was in college in the 1980s, older activists worried that young people were substituting T-shirts and political buttons for serious activism. (T-shirts were the pre-Internet hashtag.) Mere conversation and sloganeering have always seemed like a potentially dangerous sap to the energy of real protest.

The difference now is that online conversation blurs into action, in a fashion that sometimes surprises even the conversationalists. One of the curious features of online dissent is how it can emerge from discussion boards that weren't designed for political talk at all. Once you get enough people talking together online for *any* reason, they discover shared areas of social concern. As Clay Shirky documents in *Cognitive Surplus*, South Korean girls who loved the boy band Dong Bang Shin Ki would flock to the fan board Cassiopeia—about a million of them. In 2008, when the South Korean president, Lee Myung-bak, decided to allow American beef back into the country—it had been banned for years because of the mad-cow-disease scare—the girls started talking about it and decided they didn't like the idea. Thousands of people, half of whom were teenagers, joined candlelight protests in Cheonggyecheon Park in Seoul. The government tried to crack down, but this backfired when images of teenage girls getting beaten by police went public. Eventually the president partially apologized, imposed new beef import restrictions, and demanded the resignation of his entire cabinet.

Online talk may be cheap, but because it's public and linkable, it catalyzes multiples. As the British media critic Charlie Beckett writes, "It strikes me that social media embodies the connection between action and expression. For example, you can Tweet that you are going to a demonstration. The hashtag connects you to others, acts as an expression of your opinion and a call to action, and builds solidarity. It is democratic, efficient and endlessly variable. It is personal but it increases social capital for the movement."

In fact, evidence suggests that the more socially active people are online, the more civically active they are offline, too. A survey released in 2011 by the Pew Research Center found that 80 percent of people who use the Internet participated in an offline group or volunteer organization, compared to only 56 percent of those who weren't Internet users. People who were active in social networking were even more engaged offline—for example, 85 percent of those who used Twitter were involved in an offline organization. Even internationally, this seems to hold true. Emily Jacobi, head of Digital Democracy—an organization devoted to empowering human-rights activists—polled one hundred Burmese youth activists, dissidents, and refugees living on the Thai border. She found that the ones with Internet access were most likely to feel optimistic about their political action. "And we're talking about people who have to walk for miles to get to an Internet café just to get online," Jacobi tells me. "It isn't easy for them, but when they can connect to other people, they feel less alone, more committed."

Another concern about political speech online is the problem we saw in ambient contact—homophily. We're most comfortable around those similar to us politically, and the Internet makes it easy to find those people. That's what multiples are all about, after all. Logically, this could produce a terrible echo chamber effect: Political partisans reinforce one another's prejudices, increasing political partisanship. As Farhad Manjoo documented in his book *True Enough*, it isn't just about opinions—like-minded groups reinforce each other's lousy, half-true, or even patently wrong facts. This is called selective exposure: we seek out and pay close attention to facts that suit what we believe, ignoring the ones that don't.

If you visit any avowedly partisan political blog and look at the comments, you'll see this in action. You'd be tempted to go, Well, no wonder the United States is so divided: The online echo chamber has utterly deformed public debate.

Except as evidence rolls in, the picture doesn't look so simple. Indeed, the behavior at those hyperpartisan political blogs may be more an exception than the rule online, when it comes to political talk. As scholars examine online discourse, they often find that the average person isn't locked into an echo chamber at all. For example, another Pew survey found very little political agreement even among friends. You'd expect, given what we know about homophily, that friends would mostly reinforce one another's views. But in reality, Pew found, barely 25 percent "always agree or mostly agree with their friends' political postings" on social networks, and amazingly, 73 percent "only sometimes" agree or *never* agree. And 38 percent of folks on social networks "have discovered through a friend's posts that his/her political beliefs were different than the user thought they were." Public thinking and ambient contact give us a richer sense of what's on our friends' minds—and what we find surprises us.

The news-sharing abilities of tools like Twitter may expand, rather than shrink, the political exposure of users—even among the most ideological folks. One study homed in on Twitter users who displayed "clear political preference," as with right-wing Twitter users following Fox News, or left-wing Twitter users following *The New York Times*. It turns out these users inadvertently got a more diverse set of news delivered via other people they followed, because their friends would tweet tidbits from the other side of the political spectrum. Indeed, 17.8 percent of the left-wing Twitter users were seeing right-wing media via retweets from people they followed, and the right-wing Twitter users saw even *more*—57.2 percent of them were seeing left-wing media. Even when they picked their online friends homophilically, it still appeared to broaden their exposure. "Indirect media exposure," the researchers concluded, "expanded the political diversity of news users obtained by a significant amount."

Things are probably even less cloistered when you step away from the big social networks like Facebook and Twitter—and dive into the teeming world of discussion boards for hobbies and cultural passions. Joseph Kahne, a professor at Mills College, ran several studies on young people's online discussions. He found two striking things: Those who participated in communities devoted to everyday interests were exposed equally to views they did and didn't agree with. There was little echo chamber at all. More important, kids involved in hobby-based online forums had higher civic involvement than peers who weren't active: They volunteered more often, raised money, and collaborated to solve social problems. Kahne tells me he suspects this is because interest-based online groups expose kids to a wider range of people than they'd normally meet: more diverse ages, politics, and literacy. If you're interested in sailing or guitars or cross-stitching, you're probably going to run into people from all walks of life. So these forums broadened the teenagers' worlds and got them comfortable dealing with strangers, a crucial skill of maturity. (Interestingly, this *wasn't* true for kids on Facebook, because Facebook doesn't encourage you to interact with strangers.)

To be clear, I'm not suggesting that online interactions are going to magically reduce the partisanship of modern politics. It's amazingly hard to change someone's mind on a big, important issue. But this evidence contradicts the notion that online communications are singularly responsible for the increased toxicity we often perceive in today's politics. Indeed, one could quite easily argue that political divisiveness (in the United States, anyway) is still mostly a product of forces in the offline world—such as deep structural problems in how Congress works, gerrymandering, the behavior of traditional one-way media, and the billions of dollars spent on think tanks and electioneering that deform political discourse.

Why are so many political pundits so convinced that the digital world is corroding public life? Possibly—and ironically—because of selective exposure. If you're on a hunt for rancorous partisanship, it's easy to find. Screaming political blogs abound, and that's where journalists go. But these sites are, for all their prominence in the media, likely a relatively small chunk of Internet conversation as a whole. Indeed, political speech of *all* forms is probably a surprisingly small fraction of what we discuss online, if it's similar to our offline lives: Robin Dunbar's analysis of everyday conversation revealed that people talked about politics a mere 2.9 percent of the time.

It's become fashionable to talk about online dissent as "leaderless." But as with collective thinking, the truth is more complicated. Political movements that don't have a clear leader still often rely on individuals fluent in the behavior of digital crowds. Like tummelers, these individuals know how to create environments where people feel excited about participating. They know how to sense the emergent activity of online groups; they create networks, then learn from them about where to go. They know that online movements work best—arguably they only work at all—when they're focused on specific goals.

Like seeking justice for a black teenager—as in the case of Trayvon Martin. On the night of February 26, 2012, the seventeen-year-old was walking back from the grocery store to a house that his family was visiting when he was fatally shot by George Zimmerman, the gated community's neighborhood watch coordinator. Remarkably, Zimmerman wasn't charged. This angered the Martin family and observers, who believed the shooting was racially motivated. But local media paid scant attention to the case—a brief report ran the next day, and a slightly longer story the day after that in the *Orlando Sentinel*—so the police faced little pressure to act.

That pressure grew online. Eleven days after the shooting, Kevin Cunningham, a thirty-one-year-old in Washington, DC, heard about Trayvon Martin via his fraternity's e-mail list. Inspired by the Arab Spring a year earlier, he went to Change.org—a Web site for setting up petitions—and crafted one calling on the police to charge Zimmerman. He shared it with his fraternity, and it was posted to their e-mail list. Within a few days, it garnered fifteen thousand signatures. In New York, a young digital-media strategist named Daniel Maree posted a YouTube video asking people to wear hoodies on a designated day to call attention to Martin's case, since the teenager had been wearing one when killed; the action became known as the Million Hoodie March. It was a catchy and potent meme, one that neatly highlighted the racial double standards in clothing: hoodies are ubiquitous, yet young black men who wear them are regarded as suspicious by police. The wear-a-hoodie meme exploded on Twitter when several celebrities tweeted about it. They also linked to the Change.org petition, which soon amassed two million signatures—making it the biggest in Change.org's history. The tsunami of online discussion caught the mainstream media's eye, and soon Trayvon Martin was a national story. Within weeks, the police arrested and charged Zimmerman. Because the goal was specific—the police must investigate this teenager's death—the online conversation produced clear, focused action.

And as with all collective thinking online, the architecture of participation matters. What makes Ushahidi maps so powerful is that they easily allow for civic microcontribution. Since it's easy for almost anyone to contribute a tiny bit of information, almost anyone does. The Haitian earthquake of 2010 was one dramatic example. When the disaster occurred, Ushahidi's developer instantly set up a Haiti map so a team of volunteers could scour blogs, social sites, and regular media looking for information useful to rescue crews—like locations of people in trouble or stations with fuel or medicine. But the lid really blew off when they set up a mobile phone number where Haitians (or anyone anywhere, really) could text in news.

Citizens began texting up to two thousand reports a day; messages in Creole were farmed out to Creole-English speakers online, who'd translate them within minutes, another form of microcontribution. The map's accuracy was astounding. When a Haitian woman in labor texted that she was bleeding out, the translators pinpointed her location within 5 decimal degrees of latitude and longitude, as the U.S. Coast Guard discovered. "When compared side by side, Ushahidi reporting and other open sources vastly outperformed 'traditional intel,'" as Craig Clarke, an intelligence analyst for the U.S. Marine Corps, has said. And these microcontributions work not just in sudden crises but even with slower-moving civic issues: Colombians have used it to collect reports of child labor, and European skiers to amass reports on potential avalanche conditions.

Collaborative mapping is, in the world of civics, a new literacy. "Having a real-time map is almost as good as having your own helicopter. A live map provides immediate situational awareness, a third dimension, and additional perspective on events unfolding in time and space," writes Patrick Meier, Ushahidi's former head of crisis mapping. And Meier notes that the maps create a sphere for public thinking, with locals posting comments in the margins.

There's also a power shift here in the nature of this citizen information gathering. Traditionally, the state kept tabs on its citizens. But this rarely worked in reverse. Journalists provided a watchdog role, but individual citizens had few ways to participate. As everyday citizens become equipped with documentary tools, this is changing.

Steve Mann calls it "sousveillance." Surveillance is when authorities observe the population from above (in French, *sur*). Sousveillance is when the population turns its cameras on the powerful—watching from below (*sous*). It's deeply unsettling to authority figures, as Mann, a computer science professor at the University of Toronto, discovered. He'd pioneered a wearable computer in the 1980s that included a camera system that took pictures and video of what he was looking at. He intended it as a personal memory aid, but as Mann tells me, police and security guards hated it—demanding he turn it off, and if he refused, escorting him away or even tackling him.

Mann's head-mounted camera seemed crazy back then. But now that there's a camera in every phone, sousveillance has become a mass culture, with serious implications for the powerful. The sheer explosion of cameras dramatically increases the likelihood that abuses of power will be recorded. Tufekci's survey of the Tahrir Square protesters found that about 50 percent documented the protests online—so, she writes, "it becomes apparent that at least tens, if not hundreds of thousands of people were documenting the protests," producing a torrent of evidence. "There is now a constant 'price tag' on violent responses from oppressors: Somebody will catch those moments with a simple camera and upload it on YouTube," as Srdja Popović, a leader of the Serbian youth group Otpor! (which agitated for the overthrow of Slobodan Milošević in the late 1990s) told *The European*.

Do sousveillance and citizen journalism supplant traditional journalism? Not really—even the most committed citizen documentarians don't have the time or resources to do the legwork that journalists, at their best, do. Instead, it's become clear that citizens and traditional media have a symbiotic relationship. Because citizen journalists are more widespread, they capture things traditional media cannot. But their voices become amplified when they attract the attention of old media, with its audiences in traditional corridors of power. In the Iranian uprising of 2009 and Syrian uprising of 2011, the governments banned (or severely limited) international media, so citizen reports became the only path to foreign coverage; but those reports were given additional force when disseminated in the mainstream.

Tools for thinking help make people smarter. But they don't necessarily make them morally *better*. So what happens when you take brutal rulers and give them technology that makes them smarter and more efficient? They become better at doing evil.

This is what happened during the "donkey blogger" protest in Azerbaijan. In 2009, a group of young tech-savvy activists decided to embarrass their perennially corrupt and autocratic government. Flush with oil money, state officials had been caught paying eighty-two thousand euros for two donkeys imported from abroad. For the activists, this waste of money was a perfect metaphor for the government's decrepitude. So they held a funny mock press conference where one dressed in a donkey suit and the others pretended to be journalists. ("Mr. Donkey, we heard that you are much more expensive than local donkeys. What is the reason for that?" asked one young woman. "I'm better than them," the donkey replied. "I know three languages . . . and I can play on violin," he added, standing up and performing a frenetic violin solo.) The activists posted the "strangely adorable protest" (as the *Atlantic* described it) online to YouTube and other sites.

A week later, the government struck back. Police dragged two protestors, Adnan Hajizada and Emin Milli, off to prison, and sentenced them to two and two and a half years, respectively, on trumped-up charges of hooliganism. (Indeed, the government never admitted it was reacting to the video.) News of the young men's jailing traveled quickly online. While traditional media aren't very free in Azerbaijan, the government leaves the Internet relatively unfettered, at least for the well-educated, younger, and urban folks who can afford it. Since the Internet isn't very censored, the activists openly keep blogs, and they quickly relayed news of the donkey bloggers' plight. Soon, those who used social networking like Facebook and Russia's Odnoklassniki were chatting about it, too.

Now, if you believe that citizens talking to each other can erode pluralistic ignorance and catalyze dissent, then the Internet of Azerbaijan ought to have been bad for the government. It would have made people more angry, more likely to rise up.

But it appears the opposite happened. Katy Pearce, an assistant professor in the communications department at the University of Washington, who has worked and traveled extensively in the Caucasus region, decided to measure the effect of the donkey bloggers' conviction. She polled 1,795 Azerbaijanis in the fall of 2009, just after the trial, and a similar number a year later. She found that the online discussion actually repressed people's political urges. The more wired her respondents were, the *less* they believed political change would take place. It was precisely the reverse of what happened in Egypt, Tunisia, and Libya. "Unlike many of the countries in North Africa and the Middle East that experienced an Arab Spring, where the documentation of state crimes on social media mobilized the population, the arrest of Azerbaijani bloggers only demoralized frequent Internet users," Pearce wrote. Why?

Because, she tells me, the Azerbaijani government had done a clever judo move online. By aggressively targeting a few key activists but leaving conversation relatively open, it could ensure that all interested observers would quickly learn the grim fate of anyone who stepped out of line. Plus, state officials released their own online propaganda claiming the activists were trying to destabilize society. In Azerbaijan, which had suffered from years of economic and political chaos after the USSR's collapse, the public worries deeply about maintaining stability, and the propaganda tapped into those fears.

The end result was just what the Azerbaijan government wanted:

the demoralization of the elite, well-educated, wired youth most likely to support protest. "This is the way they function," as Milli told Pearce. "They punish some people and let everyone else watch. To say, 'This is what can happen to you.'" Hajizada, the other jailed donkey blogger, counters that the news isn't all bad. He tells me that the truly committed activists became even more committed. "I saw much more blogging, much more video blogging, much more people on Facebook. They became braver," he says. He suspects things bifurcated: "The larger public became afraid," but the die-hards "became even more stubborn and even more die-hard."

Despots, it turns out, are learning to practice what journalist Rebecca MacKinnon calls "networked authoritarianism"—the use of the Internet to consolidate power. Rather than simply ban all digital communications, they realize, why not leave it partially open? Then dissidents will engage in public thinking and networking, which is a great way to keep tabs on them. "Before the advent of social media, it took a lot of effort for repressive governments to learn about the people dissidents are associated with," as the writer Evgeny Morozov notes in *The Net Delusion*.

Despots also love the permanence of digital memory. The KGB stamped "хранить вечно" ("to be preserved forever") on the files it kept on activists. Even sousveillance can become a double-edged sword, with videos and photos of protests—shot by activists themselves—offering governments tidy documentation of precisely whom to arrest. After the 2009 Iranian election protests, the proregime *Raja News* printed thirty-eight photos on its Web site with sixty-five faces circled in red and another batch of forty-seven photos with roughly one hundred faces similarly circled, inviting supporters to crowdsource identification of dissidents. (As Morozov notes, Iranian police claim that tips they received in response to this invitation helped them arrest "at least forty people.") As the price of storing data drops, John Villasenor of the Brookings Institute writes, there's a troubling side effect to sousveillance: The state can simply store everything it can get its hands on—phone calls, texts, micro-blog posts, status updates, GPS coordinates of phones—and when it identifies a dissident, scour the trove for evidence. "Pervasive monitoring will provide what amounts to a time machine allowing authoritarian governments to perform retrospective surveillance," Villasenor notes.

The use of digital tools to consolidate power has arguably been raised to an art form by the Chinese government, which runs an Internet that is somewhat open yet precisely tuned to the ruling party's needs. China's Great Firewall keeps citizens from accessing foreign media; meanwhile, old-fashioned threats of jail time ensure that local Internet executives carefully self-censor the chatter on their services. The end result is a local cyberspace where it's fine to talk about pop culture and current events, or even vent about local corruption and the environment. But serious discussions of democracy are crushed. Indeed, the Tiananmen Square rebellion has been so thoroughly memory-holed that when a documentary crew in 2005 showed Beijing university students the iconic photo of a Chinese man facing down a tank, the youth had never seen it before.

There are subtler dangers in online politics, too. As I said, I suspect that concerns about echo chambers are overstated in countries that protect free speech and minority rights. But in countries with deep ethnic or ideological rivalry, homophilic sorting can be quite dangerous, as groups form online to egg each other on to violence. This is why it rarely makes sense to declare a particular form of media inherently "good" or "bad" for human rights. Tools may have inherent and even universal biases, but these are also influenced by the country in which they operate.

One could also argue the power of online action to foment change has limits, which are becoming obvious in the case of Egypt. The digital wave might have helped topple Mubarak, but that task had a relatively clear, singular purpose. Constructing a new government is much more complex, particularly in the face of a military intent on retaining power. Two years after the uprising, the country was still enmeshed in a struggle, occasionally bloody, between secular and religious parts of the population. Digital tools may help in dispelling pluralistic ignorance, but it's not yet clear how they can help when the problem has moved on to a new phase: creating trust and cooperation between groups with very different visions of how the country should be ruled.

In many ways, the biggest conundrum for politics in the digital age is how it moves speech into the private sector.

When civic conversation takes place on Facebook, Facebook's rules govern how civil society operates. This has already caused plenty of problems. For example, Facebook's rules don't allow for anonymous accounts or even pseudonymous ones. This cuts against the needs of many activists, who often wish to conceal their identities, for reasons both tactical (keeping themselves and their families safe) and existential: as pamphleteers throughout history have shown, anonymous voices have a lot of power. They encourage the audience to focus on the message instead of the messenger. In the Egyptian uprising, Wael Ghonim exploited both these levers by running his page pseudonymously. But when Facebook officials discovered what he was doing, they delisted the page—and poof, one of the important forums for organizing in Egypt vanished overnight. Facebook wouldn't put it back online until Ghonim arranged to transfer administrator rights to a friend in the United States and one in Egypt, who operated under their real names. The collision here is between commerce and civics. Facebook's finances rely on the company selling ads targeted to one's real-life self.

What's more, corporations obey the laws of the states in which they work. This has produced some egregious examples of digital-age firms bending to the whims of despots. In 2004, Yahoo! handed over to the Communist Party the identity of Shi Tao, a journalist who'd used Yahoo! e-mail to send a write-up of party directives to a U.S.-based democracy site. Shi Tao was jailed for a decade and subjected to brutal factory labor. Even Apple, the darling of the tech world, has censored material for the Chinese government: "On Apple's special store for the Chinese market, apps related to the Dalai Lama are censored, as is one containing information about the exiled Uighur dissident leader Rebiya Kadeer," MacKinnon writes. Perhaps worst of all are the many Western firms that sell to despotic regimes the very tools that are used to track dissidents and block content. The U.S. firm Blue Coat sold filtering technology to the autocrats of Tunisia and Syria, and Nokia Siemens sold tools that Iran used to track dissidents' mobile phones; Cisco's tech helps power the Great Firewall. Nor is this behavior limited to despotic regimes. Democratic countries are hardly immune to the temptations of tracking citizens illicitly; a study by the Electronic Frontier Foundation found, for example, that the FBI violated the law thousands of times while requesting digital information on citizens. Though the U.S. government loves to talk about the free flow of information, when the Web site Wikileaks released internal diplomatic documents and footage of the military killing civilians, politicians and pundits fulminated so ferociously that major U.S. firms like Amazon and Paypal cut off Wikileaks, probably worried about being on the wrong side of a political fight.

In 1996, writer and electronic activist John Perry Barlow proclaimed "A Declaration of the Independence of Cyberspace." Addressing old-school governments—"you weary giants of flesh and steel"—he proclaimed, "You are not welcome among us. You have no sovereignty where we gather." As it turns out, nothing of the sort was true.

So is there any way to conduct civic speech in the corporate digital sphere without running afoul of corporate rules or putting activists in danger?

It's certainly possible to make speech safer from prying eyes. Electronic free-speech groups worldwide for years have designed clever tools for encrypting communications. If citizens use them diligently, it's harder for state powers—or even corporate powers— to listen in. The free, open-source Tor software lets users surf with high levels of anonymity; encrypted e-mail and chat tools let citizens talk with less chance of corporations or governments listening in. Even sousveillance can be made safer. For example, Obscura-Cam, software created by the human rights and high-tech non-profits WITNESS and the Guardian Project, automatically obscures the face of anyone in a video so protesters can document their work while removing the identities of anyone who needs protection. (In 2012, YouTube released its own face-blurring tool.) Others have created DIY social networks that can be run by citizens themselves, like Crabgrass or Diaspora, so that no corporation is sitting in the center and observing what's going on.

Yet the truth is, if you want to talk to society at large, you need to work openly in the huge, for-profit spaces—because the whole point of large-scale public change is to reach as many people as possible. This cedes a lot of ground to the Facebooks of the world. But then again, activists in repressive regimes historically have always courted danger, because they always worked at some point in the open. When you live in a serious police state, the police already know who you are. This is what the donkey blogger Hajizada wrote when the crowds on Reddit asked him why he he hadn't carefully used crypto to hide his online activities. "Nothing can be hidden in Azerbaijan," Hajizada replied. "KGB is everywhere. You might as well go open." And working under your real name is often crucial to catalyzing support, "because if you go anonymously, no one will know you and if you get arrested no one will be there for you."

And there are some paradoxical upsides to working in corporate spaces. Sure, they're filled mostly with fluffy pop culture; Justin Bieber was a trending topic for ages on Twitter, and music videos bestride YouTube like a colossus. Yet this is exactly what can make these venues so useful for public agitation. Since they have such huge, varied crowds, they're the fastest route to accessing an enormous public. While politics rarely pierces the mass attention in those sites, when it *does* there are hundreds of millions watching. Better yet, because these corporate services are used so heavily for entertainment and silly reasons, authorities are loath to shut them down. This is what's known as the Cute Cat Theory of online services, as formulated by Ethan Zuckerman, the head of the MIT Center for Civic Media. Shut down a small Web site where only activists blog, and the mass public will neither notice nor care. Shut down YouTube for political reasons? Then you take away people's cat videos—which enrages and radicalizes the masses. And autocrats can't risk that.

We saw this dynamic at work in India in 2012, when an artist named Aseem Trivedi began posting "Cartoons Against Corruption" on his Web site. Incensed, politicians accused Trivedi of seditious activities and insulting the national emblem. (One cartoon depicted Parliament as a toilet.) So they pressured BigRock, an Indian provider of the country's DNS listings, to block the site. BigRock caved, and Trivedi's site was "disappeared" from the Internet. Trivedi responded cleverly, by moving his cartoons over to Blogger, Google's publishing platform. This put the Indian politicians in a quandary. They couldn't block Trivedi's blog without blocking *all* of the Blogger domain—and that would knock offline tons of legitimate Indian businesses and cultural sites that also use Blogger. Google's service may be filled with silly, trivial sites, but shutting it down would pit the government against millions of citizens. By moving over to Google, Trivedi was able to keep publishing. Even after the government jailed him in the fall of 2012, the cartoons stayed up.

It's probably possible to make these corporate spaces safer for civic discourse, if the political will exists. The "real" world, after all, took centuries of reform to make modern civic life possible. Before the Magna Carta, individuals had very few rights. The U.S. civil rights movement was fought in part over the rights of blacks to access private-sector spaces, like restaurants. And when corporations become so large that they effectively dominate their sphere, the United States (like many other countries) uses antitrust law to establish the rules of conduct. "Facebook is a utility; utilities get regulated," as danah boyd puts it.

What we need now, as MacKinnon and other thinkers have argued, is a new Magna Carta for the digital age—one that requires corporate providers of online speech to respect the rights of those who speak on their platforms. "No person or organization shall be deprived of the ability to connect to others without due process of law and the presumption of innocence," is the prime rule suggested by Tim Berners-Lee, the inventor of the Web. More countries worldwide (the United States included) could follow European Union officials and push for regulations requiring high-tech services to give users more control over their data—or deleting it upon request.

It might seem utopian to imagine this sort of regulation stitched together across nations. (In autocratic ones, it would be impossible.) But in democratic countries it's not inconceivable. Indeed, in addition to regulation, it's possible high-tech firms might agree to some voluntary standards. As MacKinnon points out, firms in democratic, open societies have responded to public pressure in the past. When student activists pressured apparel companies to reform their sweatshop practices in the 1990s, it produced the Fair Labor Association—a very imperfect solution, but one that improved conditions at some overseas factories. After Microsoft, Google, and

Yahoo! were humiliated in Congress for their cooperation with the Chinese government, they joined the Global Network Initiative, adopting voluntary rules to address the human rights implications of their work. (It's had concrete, positive effects: When Yahoo! later rolled out services in Vietnam, it assessed the country's rights record and opted to locate its servers in Singapore, where they'd be out of reach of the dissent-crushing Vietnamese government.) The challenge, then, is to get people to care about their digital rights enough for governments, and companies, to respond to them. Here, too, there are glimmers of hope, in the way that citizens in Europe and the United States, aided by high-tech firms, have successfully fought ham-fisted legislation designed to let copyright holders knock Web sites and users they don't like offline.

As Cory Doctorow points out, the great gift of the Internet is in rebalancing the stakes. States have long been able to track citizens and organize armies. Today's high tech lets them do this with much greater efficiency. But for citizens, the change has been far greater. "Every human endeavour that requires more than one person's effort has to devote a certain amount of resources to the problem of coordination: The Internet has greatly simplified this problem (think again of the hours activists used to spend simply addressing postcards with information about an upcoming demonstration)," Doctorow writes. "In so doing, it has provided a disproportionate benefit to dissidents and outsiders (who, by definition, have fewer resources to start with) than it has to the incumbent and powerful (who, by definition, have amassed enough power to squander some of it on coordination and still have enough left over to rule)."

There's a lot we don't know about the civic impact of our digital lives. Ambient awareness of our fellow citizens can help marshal protest to respond to crises. It can help us change what's wrong. Can it also help build what's right?

Epilogue_

The *Jeopardy!* clue popped up on the screen in front of me, and the host began to read the question:

"'Dynam' is a combining form meaning this five-letter word that can precede drill, broker, or brake."

My mind raced as I puzzled over the clue. Is it . . . "dynamo"? I wondered. What's a combining form, anyway? Argh, I'm a science journalist, I should know this! Maybe it's . . .

Too late. My opponent beat me to the buzzer and nailed the clue perfectly:

"What is 'power'?"

I looked in despair over at the black speaker a few feet from me, from which the robotic voice had spoken. I was playing a round of the TV trivia game *Jeopardy!*, in which the contestants race to figure out what the clue means. But I wasn't playing against a human. I was playing against Watson, a supercomputer created by IBM.

And I was being crushed. Over the next fifteen minutes we played two rounds, and I beat the computer on only about three or four clues. It nailed every other one perfectly and much faster than me, usually hitting the buzzer while I was still furrowing my brow.

Clue: "The future is inevitable, at least according to this phrase translated from the Latin 'iacta alea est.'"

Watson: "What is 'The die is cast'?"

Correct!

Clue: "This Arizona senator's send-off in 1998 was presided over by Reverend Carlozzi and Rabbi Plotkin."

Watson: "Who is Barry Goldwater?"

Correct!

Clue: "A nineteenth-century shut-in / We really don't mean to butt in / Her 'A Route of Evanescence' / Would've thrilled Donald Pleasence."

Me [*thinking*]: This is insane.

Watson: "Who is Emily Dickinson?"

Correct!

You get the picture.

Watson is IBM's high-profile sequel to Deep Blue, the chess-playing computer. Whereas Deep Blue beat the world's leading chess player, Watson was designed to do something even more impressive: It was going to play *Jeopardy!* on TV against the game's champions. While chess relies on pure logic and math, the skills required to win *Jeopardy!* are significantly weirder, as well as harder to replicate in code. To win, Watson would need to understand "natural language." It would have to possess an enormous store of knowledge about pop culture, history, sports, jokes. (Watson wasn't connected to the Internet; it only "knew" information that was in its electronic database, just as contestants on the TV show can rely only on what's in their brains.) To "teach" Watson, IBM had spent three years patiently feeding the supercomputer millions of pages' worth of documents. After my humiliating defeat at Watson's hands, I went for lunch with David Ferrucci, the witty, frenetic head of the Watson project. His creation, he said, had been trained on "books, reference material, any sort of dictionary, thesauri, folksonomies, taxonomies, encyclopedias, any kind of reference material you can imagine getting your hands on or licensing. Novels, Bibles, plays." Far more than I or any human could ever absorb, certainly.

But the spookiest thing about Watson was its ability to grapple with the clever, often obtuse wordplay for which *Jeopardy!* is famous. For example, the game has a category named "Edible Rhyme Time." A typical clue: "A long, tiresome speech delivered by a frothy pie topping." The answer is "meringue harangue," and yes, Watson could nail that, too.

Traditionally, computers have been terrible at understanding human language. That's because computer scientists have tried to teach them the rules of English (or Russian, or whatever language they were working in). This technique never worked, because language—and human knowledge—is so ambiguous that it doesn't really break down into rules very well. While humans easily absorb the ambiguous, playful aspects of language, computers don't.

Instead, Ferrucci and his team decided to harness what computers are good at: brute-force statistics. Ferrucci's group programmed Watson to analyze texts about any subject and identify what types of words are, statistically speaking, most (and least) associated with it. So, for example, when Watson saw that odd little clue for Emily Dickinson, it looked at the various bits of info in it—the title "A Route of Evanescence," the reference to a "shut-in," and even the category ("Rotten Poetry about Good Poets")—and found that these were all highly correlated with Emily Dickinson herself. It also would likely have found, for example, that "Donald Pleasance" was not well correlated with the clue and thus wasn't the answer. Watson had hundreds of techniques for analyzing and comparing texts inside its digital store, including rhyming dictionaries and collections of synonyms.

This is more than googling. Google finds you a page, but you have to read it to find the factoid you're looking for. Watson is, as Ferrucci put it, not a search engine but a "question-answering machine."

"One of the things I like to refer to is the computer on *Star Trek*," he said. "It understands what you're asking and provides just the right chunk of response that you needed." When, he asks, will we get to the point where everyday computers can talk to us? "That's *my* question."

When Watson was finally turned loose, it decimated its human opponents. In a bout broadcast on ABC, it even bested Ken Jennings, the world's leading human player, who, years before, had beaten seventy-four people in a row. Watching the machine deftly handle puns was breathtaking, even a little unnerving. It was like watching one of those Rise-of-the-Machines movies in which the computer develops self-awareness and decides to slaughter humanity just for the heck of it. (It didn't help that Watson's mechanized voice sounded like some upbeat cousin of the military computer that nearly launches a nuclear war in the movie *WarGames*.)

At one point during my time visiting IBM, I wandered into the room of slablike servers that ran the Watson software; Watson's body, as it were. I was buffeted with cold gusts, because air conditioners were running full tilt to counteract the prodigious amount of heat generated during that lightning-fast fact surfing. Watson moves at up to eighty teraflops, which is probably more than ten thousand times faster than your personal computer. When looking for answers, it parses 200 million pages of information—about equal to 2,000 books—in three seconds.

No wonder I lost.

I'll admit it: Watson scared the living heck out of me.

As I watched it deftly intuit puns, juggle millennia's worth of information, and tear through human opponents like some sort of bionic brain-shark, a technological dystopia unfolded in my grim imagination. This is it! We're finally doomed. I could imagine Watson-like AI slowly colonizing our world. Knowledge workers would be tossed out of their jobs. Shows like *Jeopardy!*—indeed, all feats of human intellectual legerdemain—would become irrelevant. And as we all started walking around with copies of Watson in our wearable computers, we'd get mentally lazier, relying on it to transactively retrieve every piece of knowledge, internalizing nothing, our minds echoing like empty hallways. Deep conversation would grind to a halt as we became ever more entranced with pulling celebrity trivia out of Watson, and eventually humans would devolve into a sort of meatspace buffer through which instances of Watson would basically just talk to *each other*. Doomed, I tell you.

Obviously, I don't really believe this dire scenario. But even for me, such dystopian predictions are easy to generate. Why is that? Among other things, doomsaying is emotionally self-protective: If you complain that today's technology is wrecking the culture, you can tell yourself you're a gimlet-eyed critic who isn't hoodwinked by high-tech trends and silly popular activities like social networking. You seem like someone who has a richer, deeper appreciation for the past and who stands above the triviality of today's life. Indeed, some clever experiments by Harvard's Teresa Amabile and others have found that when people hear negative, critical views, they regard them as inherently more intelligent than optimistic ones; when we're trying to seem smart to others, we tend to say critical, negative things. I suspect this is partly why so much high-brow tech punditry proceeds instinctively from the pessimistic view, and why we're told so often and so vehemently that today's thinking tools make us only shallow and narcissistic, that we ought to be ashamed for using them.

But as I've argued, this reflexively dystopian view is just as misleading as the giddy boosterism of Silicon Valley. Its nostalgia is false; it pretends these cultural prophecies of doom are somehow new and haven't occurred with metronomic regularity, and in nearly identical form, for centuries. And it ignores the many brilliant new ways we've harnessed new technologies, from the delightful and everyday (using funny hashtags to joke around with like-minded strangers worldwide) to the rare and august (collaborating on Fold.it to solve medical problems.)

Understanding how to use new tools for thought requires not just a critical eye, but curiosity and experimentation. That's how Garry Kasparov grappled with the meaning of Deep Blue. After the inevitable "Oh crap!" reaction, he recognized that the alien style of the machine was useful—that you could use it to create a centaur, letting the human continue to do what the human does uniquely.

This is likely how we'll work with Watson, too. As I write, IBM has begun to roll it out in the real world. Working with Memorial Sloan-Kettering Cancer Center, IBM is creating a medical version of Watson, a program that will be fed the new medical research and case studies that come out every year—so much data that no individual human doctor could ever absorb it. "A Watson, MD, if you will," as John Kelly, the mustachioed head of IBM's research labs, tells me. He envisions a doctor examining a patient and asking Watson specific questions: If this patient is on drug X and has a light fever but rapid breathing and chest pain, what might be going on? Because Watson thinks statistically, it'll give several answers, ranked from most confident to least. This will give doctors, Kelly notes, a way to rapidly tap into "the new procedures, the new medicines, the new capability that is being generated faster than physicians can absorb it on the front lines and it can be deployed." The doctors will do what doctors do best, the machines what they do best. The Sloan-Kettering experts will still need to use human intuition and experience to make the final diagnosis. But Watson will become, in effect, "the smartest medical student we have ever had," as Larry Norton, a Sloan-Kettering oncologist, puts it.

Indeed, Watson may often simply confirm what a doctor already intuits is wrong, but knowing you've queried all world medical literature gives a diagnostician added confidence. And medicine is only the first sphere where we'll see Watsonian technology emerge. You could imagine it being deployed in any area where there are time limits—sometimes urgent ones—as well as galaxies of material to consult. Watson could be useful for lawyers poring through case law or business owners parsing government regulations.

It won't be long—Kelly estimates just over ten years—before Watson-style AI can run on everyday devices like laptops, at which point we'll *all* be using it. What might that be like? Well, it's certain that we'll harness Watson as a new form of transactive memory, pinging it regularly. A Pew study found that 22 percent of all TV watchers with smartphones already use them to fact-check what they're seeing; imagine watching the news with that sort of high-level AI at your side. Or imagine if you could consult not just facts but the world's public thinking. You'd find new ways to tap into collaborative thought, locating far-flung multiples, more people talking and obsessing and arguing over the same things you are.

Frankly, I suspect that this is only the first stage. We're not thinking big enough or weird enough. A tool's most transformative uses generally take us by surprise. For example, when the mobile phone was expensive and in the hands of the few, it was purely a corporate mechanism; when the price dropped and everyone had one in their back pocket, it became a vehicle for sousveillance, for self-organizing bar crawls, for inventing rebuslike short forms of expression, for staying in ambient contact with the world. Technological habits that seem laughable (sending smiley-bedecked text messages; making LOLcats) turn out to also be world-changing (sending text messages to Ushahidi to manage a crisis; using captioned pictures to outwit Chinese censors). We'll truly figure out what Watson is for only when people begin using the software to make jokes, to play games, to hassle each other. Or to seek not answers but questions: Andy Hickl, an AI inventor, once developed a contradiction engine that took any statement and found contradictory evidence.

Imagine Watson as a way to reveal what you're *not* thinking about—what you're ignoring.

As with all new tools, we'll also have to negotiate how not to use it. Like Thad Starner with his wearable or like the librarians training their students on the biases of Google, we'll have to become literate in the biases of a Watson AI—the gravitational pull it exerts on our thought, the modes of thought it precludes, the corporate logic it inserts into everyday life. That's our challenge. We have to interrogate our most destabilizing new technologies and be aware of their dangers (economic, political, and social), to flat-out avoid the tools we find harmful—yet not be blinded to the ones that truly augment our thought and bring intellectual joy.

Are we up to that task? Tools like Watson are awfully complex, and as with Google, the way the software works is an intellectual-property secret. This is dangerous, cognitively speaking. What if you relied on Watson to answer a question and it got things wrong? How would you know it was wrong? Are we, the humans, able to critique and understand Watson well enough?

Back when I was watching Watson play *Jeopardy!* against human opponents, I struggled to follow its reasoning. This was particularly true when Watson tossed off what seemed to be nonsensical, off-base answers. For example, during one round of *Jeopardy!* the clue was to complete the phrase "Toto, I've a feeling we're not in Ka—." Watson's top guess was "not in Kansas anymore." The machine had correctly inferred that this clue referred to Dorothy's line in *The Wizard of Oz*. That wasn't surprising, of course. I pretty much expected Watson to handle a clue as simple as that.

But Watson's second, third, and fourth guesses, the ones that Watson had less confidence in, were off base in ways that struck me as bizarre. One such guess was "Steve Porcaro," the keyboardist for the band Toto. Granted, it made a vague sort of sideways sense. The clue mentioned Toto, so Watson was probably just rummaging around to find any and all things associated with that name. But Watson's next guess was . . . "Jackie Chan." Jackie Chan, the kung fu actor? What the heck did he have to do with *The Wizard of Oz*?

I wrote a story about Watson for *The New York Times Magazine* and I described this Jackie Chan anecdote with amusement. I used it as an example of how Watson, for all its supercomputer data-crunching power, could misunderstand a clue.

But I was wrong. It turns out that Watson had a good reason for considering that answer. And it was networked humans online who figured it out for me.

After I published my article, the story got posted on *The New York Times*'s Web site, and I tweeted a link. Within minutes, some of my six-thousand-odd followers had read the piece and were debating it. Meanwhile, thousands more were reading it on the *Times* site, and scores of comments were piling up in the comments thread there.

The next day, I discovered that the online crowd had deduced precisely why Watson had guessed "Jackie Chan." As one of the commenters at the site explained:

> The reason Watson derived Jackie Chan as a possible answer to the *Wizard of Oz* question comes from the *Rush Hour* franchise. In the original, James Carter (Chris Tucker) is charged with the task of babysitting Inspector Lee (Chan). Lee tries to coerce Carter into assisting him with his investigation but Carter replies with "This is the United States of James Carter now. I'm the president, I'm the emperor, I'm the king. I'm Michael Jackson; you Tito."
>
> In *Rush Hour 2* the story begins in Hong Kong, and Lee tries to turn this phrase on Carter indicating that he's now in charge. He gaffes, however, by saying that he

is Michael Jackson and Carter is Toto. Carter replies that Lee means Tito and that Toto is what they had for dinner the previous night.

Q.E.D.! Some Twitter users, it turns out, came up with the same analysis; some reached out to me via instant messaging to point out a link to a 2004 discussion-board thread of "Favorite Movie Quotes," where a *Rush Hour* fan had transcribed that precise bit of dialogue. Meanwhile, over at the *Times* site, that comment explaining the "Toto" answer was given so many positive votes that it moved near the top of the list of comments recommended by readers.

Sure, *I* had been baffled by Watson's answer. But given enough eyes, and the ability to connect with other people, some of the gnarliest puzzles can be solved. I made the mistake of thinking that just because I couldn't follow Watson's logic myself, *nobody* could. But the Internet has enabled us to exercise new powers, new ways to talk to each other, to pass things around, to quickly broadcast a hunch—"hey, wasn't there a joke involving Toto in that movie?"— and get feedback. Using the mammoth stores of knowledge online, people were able to scrutinize and pick apart the reasoning of the world's most sophisticated artificial intelligence.

For serious fun, imagine if we emulated Kasparov fully here and used Watson to create an entirely new game show—a sort of *Advanced Jeopardy!*, as it were. Imagine two teams facing one another, a human paired with a Watson-style AI. What type of fiendish, seemingly impossible clues could you throw at a cyborg composed of Ken Jennings and Watson—the brute force of a machine paired with the fuzzy intuition of a human? What puzzles could that combination tackle in everyday life?

How should you respond when you get powerful new tools for finding answers?

Think of harder questions.

Acknowledgments

This book would not exist without my agent Suzanne Gluck, who believed from the very beginning in the crazy scope of this project. I'm also indebted to Colin Dickerman, my editor at Penguin, for his superb guiding force in helping to shape my arguments.

Writing a book like this can be solitary work, but reporting it—that's social. I owe special thanks to the scores of people I interviewed for this book, each of whom took time out of their busy days to talk. As generations of journalists before me have said, you can't figure out how the world works from sitting behind your desk. I'm grateful to everyone who gave me a glimpse into their lives.

Behind the scenes, there are also dozens more who offered ideas, conversation, and feedback that shaped my work. That includes Tricia Wang, An Xiao Mina, Debbie Chachra, Liz Lawley, Zeynep Tufekci, Clay Shirky, Brooke Gladstone, Tom Igoe, Max Whitney, Terri Senft, Misha Tepper, Fred Kaplan, Howard Rheingold, danah boyd, Liz Lawley, Nick Bilton, Gary Marcus, Heidi Siwak, Ann Blair, Eli Pariser, Ethan Zuckerman, Ian Bogost, Fred Benenson, Heather Gold, Douglas Rushkoff, Rebecca MacKinnon, Cory Menscher, Mark Belinsky, Quinn Norton, Anil Dash, Cathy Marshall, Elizabeth Stock, Philip Howard, Denise Hand, Robin Sloan, Tim Carmody, Don Tapscott, Steven Johnson, Kevin Kelly, Nina Khosla, Laura Fitton, Jillian York, Hilary Mason, Craig Mod, Bre Pettis, Glenn Kelman, Susan Cain, Noah Schachtman, Irin Carmon, Matthew Battles, Cathy Davidson, Linda Stone, Jess Kimball,

Phil Libin, Kati London, Jim Marggraff, Dan Zalewski, Sasha Nemecek, Laura Miller, Brian McNely, Duncan Watts, Kenyatta Cheese, Nora Abousteit, Deanna Zandt, David Wallis, Nick Denton, Alissa Quart, Stan James, Andrew Hearst, Gary Stager, Evan Selinger, Steven Demmler, and Vint Cerf. I'm grateful to Nicholas Carr for pushing forward my thinking about memory and creativity in *The Shallows*. More than a decade ago, Carl Goodman and Rochelle Slovin from the American Museum of the Moving Image first inspired me to think about the role of moving image in our thought. My apologies to the many colleagues I've inadvertently left out here; human memory being, as I've written, rather fragile, this is a necessarily incomplete list.

For fifteen years, I've been writing magazine articles about how technology shapes our lives, and some of that work appears in this book. I'm indebted to the many editors who've given me the opportunity to explore, often at great length, my interest in this field. At *The New York Times* magazine I've been fortunate to work with Dean Robinson, Paul Tough, Gerry Mazorati, and Hugo Lindgren. Over at *Wired*, I've been lucky to be edited by Mark McClusky, Adam Rogers, Chris Anderson, Robert Capps, Adam Fisher, Evan Hansen, and Mark Horowitz. Will Bourne offered space and encouragement for my work at *Fast Company*, Chris Suellentrop and Jacob Weisberg at *Slate*, Chris Lehmann at *Bookforum*, and Sean Stanleigh and Trish Wilson at *The Report on Business Magazine*.

I first learned about public thinking when I began my blog, *Collision Detection*, back in 2002, and a small army of thoughtful commenters showed up. I couldn't list all of them here even if I tried, but their conversation has been thrilling and edifying. The same goes for my followers on Twitter and Instagram, who astonish me daily with their wit, intelligence, and helpfulness.

Any factual errors in this book are my responsibility. But I had terrific help in reducing their number from my fact-checking team:

Boris Fishman, Rob Liguori, Molly Langmuir, Thayer McClanahan, and Sean Cooper. Many thanks also go to my copy editor Greg Villepique, who has a sharp eye for detail and a wonderful ear for language.

I've been fortunate to have friends who offered not only encouragement but, over the years, inspired me with their insights into technology and society. That includes all the Wack discussion-board folks from the BBS Echo; up in Toronto, my many friends and colleagues from Ryerson, the University of Toronto, *THIS Magazine*, and *Shift*.

I've also been lucky to have a mother, sisters, and late father who have long supported my nerdy and writerly obsessions; I couldn't have done this without you.

Most important, though, was my wife, Emily. She was invaluable every step of the way—hashing out these ideas in late-night conversations, talking me down from various ledges, and giving me brilliant feedback as the first reader of these pages.

Finally, I owe a debt of thanks to my late friend Greg Sewell, who'd been urging me to write this book for ages. He guided my thinking for two decades; I wish he were here to read this.

Notes

Chapter 1: The Rise of the Centaurs

1 *In the eighteenth century, Wolfgang von Kempelen caused a stir:* Tom Standage, *The Turk: The Life and Times of the Famous Eighteenth-Century Chess-Playing Machine* (New York: Walker Publishing Company, 2002), 22–28, 102–19, 194–221.

1 *In 1915, a Spanish inventor unveiled a genuine, honest-to-goodness robot:* "Torres and His Remarkable Automatic Devices," *Scientific American Supplement*, No. 2079 (November 6, 1915): 296–98.

1 *Faced with a machine that could calculate two hundred million positions a second:* "Deep Blue," IBM, accessed March 19, 2013, www-03.ibm.com/ibm/history/ibm 100/us/en/icons/deepblue/.

2 *"I lost my fighting spirit":* Bruce Weber, "Swift and Slashing, Computer Topples Kasparov," *The New York Times*, May 12, 1997, accessed March 19, 2013, www.nytimes.com/1997/05/12/nyregion/swift-and-slashing-computer-topples-kasparov.html.

2 *"emptied completely":* Frederic Friedel, "Garry Kasparov vs. Deep Blue," *ChessBase*, May 1997, accessed March 19, 2013, www.chessbase.com/columns/column.asp?pid=146.

2 *"The Brain's Last Stand":* *Newsweek*, May 5, 1997.

2 *Doomsayers predicted that chess itself was over:* Garry Kasparov, "The Chess Master and the Computer," *The New York Review of Books*, February 11, 2010, accessed March 19, 2013, www.nybooks.com/articles/archives/2010/feb/11/the-chess-master -and-the-computer/.

2 *Then Kasparov did something unexpected:* Portions of my writing on advanced chess appeared in Clive Thompson, "The Cyborg Advantage," *Wired*, April 2010, accessed March 21, 2013, www.wired.com/magazine/2010/03/st_thompson_cyborgs/.

2 *Chess grand masters had predicted for years:* Kasparov, "The Chess Master and the Computer."

2 *Human chess players learn by spending years studying:* Garry Kasparov, *How Life Imitates Chess: Making the Right Moves, from the Board to the Boardroom* (New York: Bloomsbury, 2010), Kindle edition.

3 *If you go eight moves out in a game of chess:* Kasparov, "The Chess Master and the Computer."

3 *"One, the best one":* Diego Rasskin-Gutman, *Chess Metaphors: Artificial Intelligence and the Human Mind*, trans. Deborah Klosky (Cambridge, MA: MIT Press, 2009), 50.

3 *Together, they would form what chess players later called a centaur . . . fought Kasparov to a 3–3 draw:* Kasparov, *How Life Imitates Chess*, Kindle edition.

4 *In 2005, there was a "freestyle" chess tournament:* My account of the 2005 "freestyle" chess tournament comes from personal interviews with Steven Cramton and Zackary Stephen, as well as these reports: Kasparov, *How Life Imitates Chess*; Kasparov, "The Chess Master and the Computer"; Steven Cramton and Zackary Stephen, "The Dark Horse Theory," *Chess Horizons*, October–December 2005, 17–20, 40, accessed March 19, 2013, masschess.org/Chess_Horizons/Articles/2005-10_sample.pdf; "PAL / CSS report from the dark horse's mouth," *ChessBase*, June 6, 2005, accessed March 19, 2013, en.chessbase.com/home/TabId/211/PostId/4002467.

5 *Hydra, the most powerful chess computer in existence:* My description of Hydra comes from an e-mail interview with Hydra's creator, Chrilly Donninger, and the following reports: Tom Mueller, "Your Move," *The New Yorker*, December 12, 2005, 62–69; Jon Speelman, "Chess," *The Observer* (UK), July 3, 2005, 19; Dylan Loeb McClain, "In Chess, Masters Again Fight Machines," *The New York Times*, June 21, 2005, accessed March 19, 2013, www.nytimes.com/2005/06/21/arts/21mast.html.

5 *with software you could buy for sixty dollars:* Mueller, "Your Move," 65.

6 *The "extended mind" theory of cognition . . . short-term memory:* I owe much of my discussion of the extended mind to the work of Andy Clark, specifically *Natural-Born Cyborgs: Minds, Technologies, and the Future of Human Intelligence* (New York: Oxford University Press, 2003); *Supersizing the Mind: Embodiment, Action, and Cognitive Extension* (New York: Oxford University Press, 2008); and a personal interview with the author.

6 *"These resources enable us to pursue":* Andy Clark, "Magic Words: How Language Augments Human Computation," in *Language and Thought: Interdisciplinary Themes*, eds. Peter Carruthers and Jill Boucher (Cambridge, UK: Cambridge University Press, 1998), 173.

7 *"I actually did the work on the paper":* Clark, *Supersizing the Mind,* xxv.

7 *The printed word helped make our cognition linear and abstract:* Walter Ong, *Orality and Literacy* (New York: Routledge, 2010). I'm thinking specifically about his argument in the chapter "Writing Restructures Consciousness," 77–114.

7 *"It is impossible that old prejudices and hostilities should longer exist":* Charles F. Briggs and Augustus Maverick, *The Story of the Telegraph, and a History of the Great Atlantic Cable* (New York: Rudd & Carleton, 1858), 22. Accessed March 19, 2013, via Google Books, books.google.com/books?id=NFVHAAAAIAAJ.

7 *"We are eager to tunnel under the Atlantic":* Henry D. Thoreau, *Walden: A Fully Annotated Edition*, ed. Jeffrey S. Cramer (New Haven, CT: Yale University Press, 2004), 50–51.

8 *called this the bias of a new tool:* Harold Innis, *The Bias of Communication*, introduction by Alexander John Watson (Toronto: University of Toronto Press, 2008).

9 *"tools for thought":* Rheingold introduced the term in his 1985 book of that title, reprinted in 2000: Howard Rheingold, *Tools for Thought: The History and Future of Mind-Expanding Technology* (Cambridge, MA: MIT Press, 2000).

9 *"The future is already here":* Garson O'Toole, "The Future Has Arrived—It's Just Not Evenly Distributed Yet," *Quote Investigator*, January 24, 2012, accessed March 19, 2013, quoteinvestigator.com/2012/01/24/future-has-arrived/.

12 *"that horrible mass of books which keeps on growing . . . return to barbarism"*: Blair, *Too Much to Know*, Kindle edition.

12 *In the sixteenth century, humanity faced:* My analysis of early publishing here owes much to Ann Blair's book *Too Much to Know: Managing Scholarly Information before the Modern Age* (New Haven, CT: Yale University Press, 2011), Kindle edition, and a personal interview with the author.

13 *"I'm not thinking the way I used to think"*: Nicholas Carr, *The Shallows* (New York: Norton, 2010), 5.

13 *precisely why societies have engineered massive social institutions:* I owe this point to Stephen Pinker, "Mind over Mass Media," *The New York Times*, June 10, 2010, last accessed March 19, 2013, www.nytimes.com/2010/06/11/opinion/11Pinker. html.

14 *"There will eventually be neuroscientific explanations"*: Gary Marcus, "Neuroscience Fiction," *News Desk* (a *New Yorker* blog), Dec. 2, 2012, accessed March 19, 2013, www.newyorker.com/online/blogs/newsdesk/2012/12/what-neuroscience-really -teaches-us-and-what-it-doesnt.html.

14 *a single brain-scanning study that probed how people's brains respond to using the Web:* Carr, *The Shallows*, 120–22.

14 *teams of Yale and University of Michigan scientists:* This material comes from James E. Swain, "The Human Parental Brain: In Vivo Neuroimaging," *Progress in Neuro-Psychopharmacology & Biological Psychiatry* 35, no. 5 (July 2011): 1242–54; Swain et al., "Functional Neuroimaging and Psychology of Parent-Infant Attach-ment in the Early Postpartum," *Annals of General Psychiatry*, 2006, 5 (Suppl 1): S85, doi:10.1186/1744-859X-5-S1-S85; and a personal interview with Swain. His work is also described in Anna Abramson and Dawn Rouse, "The Postpartum Brain," *Greater Good* (Spring 2008), accessed March 19, 2013, greatergood.berke ley.edu/article/item/postpartum_brain/.

15 *living in cities . . . stresses us out on a straightforwardly physiological level:* Clive Thompson, "The Ecology of Stress," *New York*, January 24, 2005, accessed March 19, 2013, nymag.com/nymetro/urban/features/stress/10888/.

15 *makes us 50 percent more productive:* Edward Glaeser, *Triumph of the City: How Our Greatest Invention Makes Us Richer, Smarter, Greener, Healthier, and Hap-pier* (New York: Penguin, 2011), Kindle edition.

15 *"the city's edge in producing ideas"*: Ibid.

15 *"Density has costs as well as benefits"*: Ibid.

16 *in June 2011, chess master Christoph Natsidis was caught:* Dylan Loeb McClain, "Dis-qualified by Evidence on a Phone," *The New York Times*, June 12, 2011, accessed March 20, 2013, www.nytimes.com/2011/06/12/crosswords/chess/chess-christoph -natsidis-punished-for-cheating.html; Peter Doggers, "Player Caught Cheating at German Championship," *ChessVibes* (blog), June 05, 2011, accessed March 20, 2013, www.chessvibes.com/reports/player-caught-cheating-at-german-champion ship.

17 *learning grand-master-level chess was a slow, arduous affair . . . age of twelve:* This material is from a personal interview with Frederic Friedel, and Kasparov, "The Chess Master and the Computer."

Chapter 2: We, the Memorious

20 *When one of Roy's grad students analyzed this slowdown:* Brandon Cain Roy, "The Birth of a Word" (PhD dissertation, MIT, 2013), 93–102, accessed March 21, 2013, www.media.mit.edu/cogmac/publications/bcroy_thesis_FINAL.pdf.

20 *"It's as if he shifted his cognitive effort":* Deb Roy, "The Birth of a Word," *Huffington Post*, February 8, 2013, accessed March 21, 2013, hwww.huffingtonpost.com/deb -roy/the-birth-of-a-word_b_2639625.html.

20 *Another grad student discovered that the boy's caregivers:* Matthew Miller, "Semantic Spaces: Behavior, Language and Word Learning in the Human Speechome Corpus" (master's thesis, MIT, 2011), 92–93, accessed March 21, 2013, www.media.mit.edu /cogmac/publications/mmiller_MS_thesis.pdf.

20 *It's a tantalizing finding, Roy points out:* Deb Roy, "The Birth of a Word."

23 *"I confess I do not believe in time":* Vladimir Nabokov, *Speak, Memory: An Autobiography Revisited* (New York: Vintage, 2011), Kindle edition.

23 *"The past is never dead. It's not even past":* William Faulkner, *Requiem for a Nun* (New York: Vintage, 2011), Kindle edition.

24 *"It's over, isn't it?" one of Perry's donors asked:* Jeff Zeleny and Ashley Parker, "After 'Oops' in Debate, Perry Says He'll Stay in Race," *The New York Times*, November 9, 2011, accessed March 21, 2013, www.nytimes.com/2011/11/10/us/politics/perry -gaffe-support-for-cain-at-republican-debate.html.

24 *German psychologist Hermann Ebbinghaus ran a long, fascinating experiment on himself:* Hermann Ebbinghaus, *Memory: A Contribution to Experimental Psychology*, trans. Henry A. Ruger and Clara E. Bussenius (New York: Evergreen Review, 2009), Kindle edition.

25 *In the 1970s and '80s, the psychologist Willem Wagenaar tried something a bit more true to life:* Willem A. Wagenaar, "My Memory: A Study of Autobiographical Memory over Six Years," *Cognitive Psychology* 18, no. 2 (April 1986): 225–52.

26 *Memory isn't passive; it's active:* My general descriptions of memory throughout this chapter owe much to Daniel Schacter, *The Seven Sins of Memory: How the Mind Forgets and Remembers* (New York: Houghton Mifflin, 2001), as well as a personal interview with the author; Joshua Foer, *Moonwalking with Einstein: The Art and Science of Remembering Everything* (New York: Penguin, 2011); and Gillian Cohen and Martin A. Conway, eds., *Memory in the Real World* (New York: Psychology Press, 2008).

27 *in what's known as reconsolidation:* Karim Nader, "Memory Traces Unbound," *Trends in Neurosciences* 26, no. 2 (February 2003): 65–72, accessed March 21, 2013, www.neuro.iastate.edu/Uploads/Nader_TINS2003.pdf; Almut Hupbach, Rebecca Gomez, and Lynn Nadel, "Episodic Memory Reconsolidation: Updating or Source Confusion?" *Memory* 17, no. 5 (July 2009): 502–10; Daniela Schiller and Elizabeth A. Phelps, "Does Reconsolidation Occur in Humans?" *Frontiers in Behavioral Neuroscience* 5 (May 17, 2011), accessed March 21, 2013, www.ncbi.nlm.nih .gov/pmc/articles/PMC3099269/; Alain Brunet, Andrea R. Ashbaugh, Daniel Saumier, Marina Nelson, Roger K. Pitman, Jacques Tremblay, Pascal Roullet, and Philippe Birmes, "Does Reconsolidation Occur in Humans: A Reply," *Frontiers in Behavioral Neuroscience* 5 (October 31, 2011), accessed March 21, 2013, www .ncbi.nlm.nih.gov/pmc/articles/PMC3204461/.

27 *In 1962, the psychologist Daniel Offer asked a group:* Daniel Offer, Marjorie Kaiz, Kenneth I. Howard, and Emily S. Bennett, "The Altering of Reported Experiences," *Journal of the American Academy of Child and Adolescent Psychiatry* 39, no. 6 (June 2000), 735–42.

28 *research shows that most of us just copy our old hard drives:* Catherine C. Marshall, "Challenges and Opportunities for Personal Digital Archiving," in *I, Digital: Personal Collections in the Digital Era,* ed. Christopher A. Lee (Chicago: American Library Association, 2011), 90–114; also a personal interview with the author.

29 *Bell is probably the world's most ambitious and committed lifelogger:* Portions of my writing on Bell and other lifeloggers appeared previously in Clive Thompson, "A Head for Detail," *Fast Company,* November 2006, accessed March 21, 2103, www .fastcompany.com/58044/head-detail.

33 *Google's famous PageRank system looks at social rankings:* Sergey Brin and Lawrence Page, "The Anatomy of a Large-Scale Hypertextual Web Search Engine," *Computer Networks and ISDN Systems* 30 (1998), 107–17.

33 *"On Exactitude in Science":* Jorge Luis Borges, *Collected Fictions,* trans. Andrew Hurley (New York: Penguin, 1999), 325.

37 *Jonathan Wegener, a young computer designer who lives in Brooklyn:* Portions of my writing here on Wegener and Timehop appeared in "Clive Thompson on Memory Engineering," *Wired* (October 2011), accessed March 21, 2013, www.wired.com/ magazine/2011/09/st_thompson_memoryengineering/.

40 *In the 1920s, Russian psychologist Aleksandr Luria examined Solomon Shereshevskii:* Aleksandr R. Luria, *The Mind of a Mnemonist: A Little Book about a Vast Memory,* trans. Lynn Solotaroff (Boston: Harvard University Press, 1987), 9–13, 70, 111–36.

42 *He suggests that digital tools should be designed:* Viktor Mayer-Schönberger, *Delete: The Virtue of Forgetting in the Digital Age* (Princeton, NJ: Princeton University Press, 2011), 169–95. Parts of my writing here first appeared in my column "Clive Thompson on Remembering Not to Remember in an Age of Unlimited Memory," *Wired,* August 2009, accessed March 23, 2013, www.wired.com/techbiz/people/ magazine/17-08/st_thompson.

43 *The majority kept everything:* Jason Zalinger, "Gmail as Storyworld: How Technology Shapes Your Life Narrative" (PhD dissertation, Rensselaer Polytechnic Institute, 2011), 84, 97–98, 103, 137–38, accessed March 23, 2013, gradworks.umi. com/34/76/3476270.html. The names Zalinger cites in his dissertation are pseudonyms.

Chapter 3: Public Thinking

46 *Consider these current rough estimates:* "Email Statistics Report, 2012–2016," Radicati Group, posted April 2012, accessed March 21, 2013, www.radicati.com/wp/wp-con tent/uploads/2012/04/Email-Statistics-Report-2012-2016-Executive-Summary.pdf; Hayley Tsukayama, "Twitter Turns 7: Users Send over 400 Million Tweets per Day," *Washington Post,* March 21, 2013, accessed March 21, 2013, www.washingtonpost .com/business/technology/twitter-turns-7-users-send-over-400-million-tweets-per-day/2013/03/21/2925ef60-9222-11e2-bdea-e32ad90da239_story.html; Facebook estimates from personal e-mail communication with Facebook research scientist

Cameron Marlow; "Posting Activity," Wordpress, accessed March 21, 2013, en.word press.com/stats/posting/; Duncan Hewitt, "Weibo Brings Change to China," *BBC News Magazi ne*, July 31, 2012, accessed March 21, 2013, www.bbc.co.uk/news/mag azine-18887804; "Fascinating Facts," Library of Congress, accessed March 23, 2013, www.loc.gov/about/facts.html; Michael Minges, "Overview," in *Maximizing Mobile* (Washington, DC: World Bank, 2012), 17, accessed March 23, 2013, siteresources. worldbank.org/EXTINFORMATIONANDCOMMUNICATIONANDTECHNOL OGIES/Resources/IC4D-2012-Report.pdf. Note that the chart in this latter report, the left-hand chart, mistakenly reads "millions" instead of "billions."

48 *"Ninety percent of everything is crap":* "Full record for Sturgeon's Law n.," *Science Fiction Citations*, last modified August 5, 2010, accessed March 23, 2013, www .jessesword.com/sf/view/328.

48 *a minority of people are doing most of the creation we see online:* This phenomenon is documented frequently, including in Kristen Purcell et al., *Understanding the Par- ticipatory News Consumer* (Pew Internet & American Life Project, March 1, 2010), accessed March 21, 2013, www.pewinternet.org/Reports/2010/Online-News.aspx; Alex Cheng and Mark Evans, "An In-Depth Look inside the Twitter World," Syso- mos, June 2009, accessed March 21, 2013, www.sysomos.com/insidetwitter/; Matt Carmichael, "Stat of the Day: 63% of Readers Don't Care about Your Comments," *Ad Age*, August 19, 2011, accessed March 21, 2013, adage.com/article/adagestat/ 63-readers-care-site-comments/229341/.

49 *Research suggests that even in the United Kingdom's peak letter-writing years:* Richard Harper, *Texture: Human Expression in the Age of Communications Overload* (Cambridge, MA: MIT Press, 2010), 30–33.

49 *5.15 per year:* David M. Henkin, *The Postal Age: The Emergence of Modern Com- munications in Nineteenth-Century America* (Chicago: University of Chicago Press, 2006), Kindle edition.

50 *Literacy in North America has historically been focused on reading, not writing:* Debo- rah Brandt, *Literacy and Learning: Reflections on Writing, Reading, and Society* (New York: John Wiley & Sons, 2009); and a personal interview with the author.

51 *"People read in order to generate writing":* Brandt, *Literacy and Learning*, 157.

51 *"I do not sit down at my desk":* Alice Brand, *The Psychology of Writing: The Affective Experience* (New York: Greenwood Press, 1989), Kindle edition.

52 *"bird and lady took such possession of the scene":* Quoted in Conor Cruise O'Brien, "Passion and Cunning: An Essay on the Politics of W. B. Yeats (Excerpt)," in *Yeats's Political Identities: Selected Essays*, ed. Jonathan Allison (Ann Arbor: University of Michigan Press, 1996), 49.

52 *"It crystallizes you":* Brandt, *Literacy and Learning*, 157.

52 *"Blogging forces you to write down your arguments and assumptions":* Gabriel Wein- berg, "Why I Blog," *Gabriel Weinberg's Blog*, August 29, 2011, accessed March 21, 2013, www.gabrielweinberg.com/blog/2011/08/why-i-blog.html.

55 *a group of Vanderbilt University professors in 2008:* Bethany Rittle-Johnson, Megan Saylor, and Kathryn E. Swygert, "Learning from Explaining: Does It Matter If Mom Is Listening?"*Journal of Experimental Child Psychology* 100, no. 3 (July 2008): 215–24.

55 *When asked to write for a real audience of students in another country:* This specific finding is in Mark Ward, "Squaring the Learning Circle: Cross-Classroom Collabo-

rations and the Impact of Audience on Student Outcomes in Professional Writing," *Journal of Business and Technical Communication* 23, no. 1 (January 2009): 61–82; there are similar findings in Moshe Cohen and Margaret Riel, "The Effect of Distant Audiences on Students' Writing," *American Educational Research Journal* 26, no. 2 (Summer 1989): 143–59.

56 *"reading maketh a full man, conference a ready man, and writing an exact man":* Francis Bacon, "Of Studies," in *The Essays or Counsels, Civil and Moral, of Francis Ld. Verulam Viscount St. Albans,* Project Gutenberg, last updated January 25, 2013, accessed March 22, 2013, www.gutenberg.org/files/575/575-h/575-h.htm#link2H_4_0050.

57 *"guy sitting in his living room in his pajamas writing what he thinks":* This is the infamous phrase of Jonathan Klein, a former CBS and CNN executive, quoted in "How the Blogosphere Took on CBS' Docs," a partial transcript of *Special Report with Brit Hume,* Fox News, Sept. 17, 2004, accessed March 22, 2013, www.foxnews.com/story/0,2933,132494,00.html.

57 *Early evidence came in 1978:* Norman J. Slamecka and Peter Graf, "The Generation Effect: Delineation of a Phenomenon," *Journal of Experimental Psychology: Human Learning and Memory* 4, no. 6 (1978): 592–604.

58 *it was famously documented in 1922:* William F. Ogburn and Dorothy Thomas, "Are Inventions Inevitable? A Note on Social Evolution," *Political Science Quarterly* 37, no. 1 (March 1922): 83–98. It's worth noting that the discussion of multiples is itself, as you might expect, a multiple; many writers today who write about creativity discuss the phenomenon. Some of my favorite analyses include Kevin Kelly in *What Technology Wants* (New York: Penguin, 2010), Kindle edition; Steven Johnson, *Where Good Ideas Come From: The Natural History of Innovation* (New York: Penguin, 2010), 34–35; and Malcolm Gladwell, "In the Air," *The New Yorker,* May 12, 2008, accessed March 22, 2013, www.newyorker.com/reporting/2008/05/12/080512fa_fact_gladwell.

60 *Robert Merton took up the question of multiples:* Robert K. Merton, "Singletons and Multiples in Scientific Discovery: A Chapter in the Sociology of Science," *Proceedings of the American Philosophical Society* 105, no. 5 (October 13, 1961): 470–86. Merton discusses the mathematicians on page 479.

60 *the tragic story of Ernest Duchesne:* My account of Duchesne's discovery of penicillin comes from Serge Duckett, "Ernest Duchesne and the Concept of Fungal Antibiotic Therapy," *The Lancet* 354 (December 11, 1999): 2068–71; Kurt Link, *Understanding New, Resurgent, and Resistant Diseases: How Man and Globalization Create and Spread Illness* (Westport, CT: Greenwood Publishing Group, 2007), 13–14; Paul André, M. C. Schraefel, Jaime Teevan, and Susan T. Dumais, "Discovery Is Never by Chance: Designing for (Un)Serendipity," *Proceedings of the Seventh ACM Conference on Creativity and Cognition* (2009), 305–14.

61 *consider what happened next to Ory Okolloh:* My account of the origins of Ushahidi draws from Ory Okolloh, "Update Jan 3 11:00 pm," *Kenyan Pundit,* January 3, 2008, accessed March 22, 2013, www.kenyanpundit.com/2008/01/03/update-jan-3-445-1100-pm/; Erik Hersman, "It's Not about Us, It's about Them," *WhiteAfrican,* January 4, 2008, accessed March 22, 2013, whiteafrican.com/2008/01/04/its-not-about-us-its-about-them/; Ted Greenwald, "David Kobia, 32," *MIT Technology Review,* September 2010, www2.technologyreview.com/tr35/profile.aspx?trid=947; and an e-mail communication with Hersman.

63 *Three decades after Duchesne made his discovery of penicillin, Alexander Fleming in 1928:* Alexander Kohn, *Fortune or Failure: Missed Opportunities and Chance Discoveries* (Cambridge, MA: Blackwell, 1989), 76–96.

65 *"most of the smartest people work for someone else":* Lewis DVorkin, "Forbes Contributors Talk About Our Model for Entrepreneurial Journalism," *Forbes*, December 1, 2011, accessed March 22, 2013, www.forbes.com/sites/lewisdvorkin/2011/12/01/forbes-contributors-talk-about-our-model-for-entrepreneurial-journalism/.

65 *We can see this in the history of "giving credit" in social media:* I'm drawing the history here from Rebecca Blood, "How Blogging Software Reshapes the Online Community," *Communications of the ACM* 47, no. 12 (December 2004): 53–55; Evan Henshaw-Plath, "Origin of the @reply—Digging Through Twitter's History," *Anarchogeek* (blog), July 9, 2012, accessed March 22, 2013, anarchogeek.com/2012/07/09/origin-of-the-reply-digging-through-twitters-history/; Liz Gannes, "The Short and Illustrious History of Twitter #Hashtags," *GigaOM*, April 30, 2010, accessed March 23, 2013, gigaom.com/2010/04/30/the-short-and-illustrious-history-of-twitter-hashtags/.

66 *She found that their error rate has barely risen at all:* Andrea A. Lunsford and Karen J. Lunsford, "'Mistakes Are a Fact of Life': A National Comparative Study," *College Composition and Communication* 59, no. 4 (2008): 781–806. Some of my writing on Lunsford's work originally appeared in "Clive Thompson on the New Literacy," *Wired*, September 2009, accessed March 22, 2013, www.wired.com/techbiz/people/magazine/17-09/st_thompson.

66 *one analyzed 1.5 million words from instant messages by teens:* Sali A. Tagliamonte and Derek Denis, "Linguistic Ruin? Lol! Instant Messaging and Teen Language," *American Speech* 83, no. 1 (Spring 2008), accessed March 22, 2013, web.uvic.ca/ling/coursework/ling395/395_LOL.pdf. In her book *Always On: Language in an Online and Mobile World* (New York: Oxford University Press, 2010), Naomi Baron also advances the possiblity that our increased use of text messaging might make our overall writing standards more casual. Others, like the linguist David Crystal, in his book *Txtng: The Gr8 Db8* (New York, Oxford University Press, 2009), argue that many texting short forms are actually like rebus puzzles, which require a surprising amount of work to encode and decode, and thus are the opposite of casual and lazy.

68 *it ran a comprehensive story:* Kate Zernike, "Jury Finds Spying in Rutgers Dorm Was a Hate Crime," *The New York Times*, March 16, 2012, accessed March 22, 2013, www.nytimes.com/2012/03/17/nyregion/defendant-guilty-in-rutgers-case.html.

69 *I cannot help feeling, Phaedrus, that writing is unfortunately like painting:* Plato, *Phaedrus*, trans. Benjamin Jowett, public domain, Kindle edition.

70 *Wikipedia has already largely moved past its period of deep suspicion:* The article comparing Wikipedia and the *Encyclopedia Britannica* is Jim Giles, "Internet Encyclopaedias Go Head to Head," *Nature* 438, no. 15 (December 15, 2005), 900–901. In comparison, an assessment of Wikipedia's articles on U.S. politicians found "that Wikipedia is almost always accurate when a relevant article exists, but errors of omission are extremely frequent"; see Adam R. Brown, "Wikipedia as a Data Source for Political Scientists: Accuracy and Completeness of Coverage," *PS: Political Science and Politics* 44 (April 2011): 339–43, accessed March 22, 2013, adambrown

.info/docs/research/brown-2011-wikipedia-as-a-data-source.pdf. The Wikipedia Foundation's own study of Wikipedia's accuracy is here: Imogen Casebourne, Chris Davies, Michelle Fernandes, and Naomi Norman, "Assessing the Accuracy and Quality of Wikipedia Entries Compared to Popular Online Encyclopedias," Wikipedia Foundation, August 1, 2012, accessed March 22, 2103, commons.wikimedia .org/wiki/File:EPIC_Oxford_report.pdf.

70 *as the author David Weinberger points out:* David Weinberger, *Everything Is Miscellaneous: The Power of the New Digital Disorder* (New York: Henry Holt, 2007), 140–47.

71 *"This is historiography":* James Bridle, "On Wikipedia, Cultural Patrimony, and Historiography," booktwo.org, September 6, 2010, accessed March 22, 2013, booktwo .org/notebook/wikipedia-historiography/.

71 *"typographical fixity" of paper:* Elizabeth L. Eisenstein, *The Printing Press as an Agent of Change* (Cambridge, UK: Cambridge University Press, 1980), 113.

71 *fall prey to what psychologists call recency:* Ira Rosofsky, "Toyota Laid Low by the Recency Effect—Oy,What a Feeling!" *Psychology Today* (blog), February 5, 2010, accessed March 22, 2013, www.psychologytoday.com/blog/adventures-in-old -age/201002/toyota-laid-low-the-recency-effect-oywhat-feeling.

71 *one analysis by HP Labs:* Ethan Bauley, "HP Research Shows Mainstream Media Drive Twitter 'Trends' to a Surprising Degree," *Data Central* (HP corporate blog), February 14, 2011, accessed March 22, 2013, h30507.www3.hp.com/t5/Data-Central/HP-research-shows-mainstream-media-drive-Twitter-trends-to-a/ba -p/87985; Sitaram Asur, Bernardo A. Huberman, Gabor Szabo, and Chunyan Wang, "Trends in Social Media: Persistence and Decay," presented at the Fifth International Association for the Advancement of Artificial Intelligence Conference on Weblogs and Social Media (2011), accessed March 22, 2013, www.aaai.org/ocs/in dex.php/ICWSM/ICWSM11/paper/viewFile/2815/3205.

72 *Artificial intelligence pioneer Marvin Minsky describes human smarts:* Marvin Minsky, *The Society of Mind* (New York: Simon and Schuster, 1988) and Marvin Minsky, *The Emotion Machine: Commonsense Thinking, Artificial Intelligence, and the Future of the Human Mind* (New York: Simon and Schuster, 2007).

73 *The sheer volume of questions answered is remarkable:* "1 Billion Answers Served!" *Yahoo! Answers Blog,* May 3, 2010, accessed March 22, 2013, yanswersblog.com/ index.php/archives/2010/05/03/1-billion-answers-served/; Jiang Yang, Xiao Wei, Mark S. Ackerman, and Lada A. Adamic, "Activity Lifespan: An Analysis of User Survival Patterns in Online Knowledge Sharing Communities," presented at the Fourth International Association for the Advancement of Artificial Intelligence Conference on Weblogs and Social Media (2010), accessed March 22, 2013, misc.si .umich.edu/media/papers/ICWSM10_YangJ_10.pdf.

74 *those questions are hosted in a proprietary database:* Choe Sang-Hun, "Crowd's Wisdom Helps South Korean Search Engine Beat Google and Yahoo," *The New York Times,* July 4, 2007, accessed March 22, 2013, www.nytimes.com/2007/07/04/busi ness/worldbusiness/04iht-naver.1.6482108.html.

74 *"Who is history's greatest badass":* "Who is history's badass, and why?" Quora, accessed March 22, 2013, www.quora.com/Badassery/Who-is-historys-greatest -badass-and-why.

74 *soft-spoken cofounder of Quora:* Cheever left Quora in 2012.

75 *answering begets answering:* Kevin K. Nam, Mark S. Ackerman, and Lada A. Adamic, "Questions in, Knowledge in?: A Study of Naver's Question Answering Community," conference paper presented at Proceedings of the Twenty-seventh International Conference on Human Factors in Computing Systems (2009), accessed May 30, 2013, http://web.eecs.umich.edu/~ackerm/pub/09b51/chi09-naver.final .pdf; and Yang et al., "Activity Lifespan."

76 *"architecture of participation":* Tim O'Reilly, "The Architecture of Participation," O'Reilly.com, June 2004, accessed March 22, 2013, oreilly.com/pub/a/oreilly/tim/ articles/architecture_of_participation.html.

81 *activists began using a mobile app called Vibe:* Jenna Wortham, "Messaging App Grows with Wall Street Protests," *Bits* (*New York Times* blog), October 12, 2011, accessed March 22, 2013, bits.blogs.nytimes.com/2011/10/12/anonymous-messaging -app-vibe-gets-boost-from-occupy-wall-street/; Rachel Silver, "How to Communicate Anonymously with Vibe," Movements.org, accessed March 22, 2013, www .movements.org/how-to/entry/how-to-use-communicate-anonymously-with-vibe1.

82 *a cadre of marginaliasts becoming so well liked:* I originally reported on social marginalia, and Stein's predictions, in "Clive Thompson on the Future of Reading in a Digital World," *Wired*, May 2009, accessed March 22, 2013, www.wired.com/tech biz/people/magazine/17-06/st_thompson.

Chapter 4: The New Literacies

83 *"Abraham Lincoln riding on a vacuum cleaner":* Danny Hakim, "The Shape of Things," *The Empire Zone* (*New York Times* blog), April 12, 2007, accessed March 22, 2013, empirezone.blogs.nytimes.com/2007/04/12/the-shape-of-things/.

83 *a slender 4 percent of incumbent:* "Reshaping New York: Ending the Rigged Process of Partisan Gerrymandering with an Impartial and Independent Redistricting Process," Citizens Union Foundation, November 2011, 3–4.

88 *One blogger, Heather Williams:* Heather Williams, "The politics of Obama versus McCain using Wordle," *Heather Williams: A work in pogress*, June 20, 2008, accessed May 30, 2013, http://heatherwilliams.wordpress.com/2008/06/20/the-politics -of-obama-versus-mccain-using-wordle/.

88 *an experiment in which students input the first draft:* Melissa Baralt, Susan Pennestri, and Marie Selvandin, "Using Wordles to Teach Foreign Language Writing," *Language Learning & Technology* 15, no. 2 (June 2011): 12–22, accessed March 22, 2013, llt.msu.edu/issues/june2011/actionresearch.pdf.

90 *"Once I saw the pattern of my mood going up":* Alexandra Carmichael, "The Transformative Power of Sharing Mood," Quantified Self, June 6, 2011, accessed March 22, 2013, quantifiedself.com/2011/06/the-transformative-power-of-sharing-mood/ #more-1911.

90 *technologies that let your mobile phone sense your heart rate:* Chris Weiss, "Apple Patents Heart Sensors for iPhone Recognition," GadgetCrave.com, accessed March 22, 2013, gadgetcrave.com/apple-patents-heart-sensors-for-iphone-recognition/6960/.

91 *This produced some delightful epiphanies:* Dale Lane, "Smile!" *Dale Lane* (blog), April 3, 2012, accessed March 22, 2013, dalelane.co.uk/blog/?p=2092#more-2092; Dale Lane, "Has Today Been a Good Day?" *Dale Lane* (blog), April 16, 2012, accessed March 22, 2013, dalelane.co.uk/blog/?p=2125.

92 *analyzed the color usage in Van Gogh's major paintings:* Cory Doctorow, "Van Gogh Pie-Charts," *Boing Boing*, January 29, 2011, accessed March 23, 2013, boingboing .net/2011/01/29/van-gogh-pie-charts.html.

92 *a "sentiment analysis" of the Bible:* "Applying Sentiment Analysis to the Bible," OpenBible.info, October 10, 2011, accessed March 22, 2013, www.openbible.info/ blog/2011/10/applying-sentiment-analysis-to-the-bible/.

92 *how characters interact in* Hamlet: Richard Beck, "Hamlet and the Region of Death," *Boston Globe*, May 29, 2011, accessed March 23, 2013, www.boston.com/boston globe/ideas/articles/2011/05/29/hamlet_and_the_region_of_death/.

93 *Tufte analyzed 217 data graphics:* Edward R. Tufte, *The Cognitive Style of Power Point* (Cheshire, CT: Graphics Press, 2003), 4–5.

94 *a six-minute video to YouTube that offered an explanation:* "Breaking Bad Finale Theory—a Case for Walt Poisoning Brock," jcham979, YouTube, October 5, 2011, accessed March 23, 2013, www.youtube.com/watch?v=3BROfhjCycY; Angie Han, "Watch a Video That Explains How the Ending of the 'Breaking Bad' Season Finale Went Down," /Film, October 11, 2011, accessed March 23, 2013, www.slashfilm .com/watch-video-explains-breaking-bad-season-finale/.

95 *In his 1985 book* Amusing Ourselves to Death: Neil Postman, *Amusing Ourselves to Death: Public Discourse in the Age of Show Business*, rev. ed. (1985; New York: Penguin, 2005), Kindle edition. The irony of my having reread Postman's book in Kindle format is not lost on me.

96 *a ten-thousand-word-long discussion thread:* Alex Fletcher, "'Sherlock' and 'The Reichenbach Fall': How Did He Fake His Own Death?" *Digital Spy*, January 17, 2012, accessed March 23, 2013, www.digitalspy.com/british-tv/s129/sherlock/tubetalk/ a360476/sherlock-and-the-reichenbach-fall-how-did-he-fake-his-own-death.html.

97 *the photographer Eadweard Muybridge helped settle an issue:* Marta Braun, *Eadweard Muybridge* (London: Reaktion Books, 2010), Kindle edition; I previously wrote about Muybridge in Clive Thompson, "The Animated GIF: Still Looping after All These Years," *Wired*, December 2012, accessed March 23, 2103, www.wired.com/ underwire/2013/01/best-animated-gifs/.

97 *a "nine-foot-tall rack" of hard-disk recorders and monitors:* Brian Stelter, "Video File Puts the Teeth in Sound Bites," *The New York Times*, July 16, 2009, accessed March 23, 2103, www.nytimes.com/2009/07/19/arts/television/19stel.html.

98 *the Promise TV box:* Rupert Goodwins, "Four-Strong Team Builds Digital TV Revolution," ZDNet, March 25, 2012, accessed March 23, 2013, www.zdnet.com/four -strong-team-builds-digital-tv-revolution-3040154872/; "What is Promise.tv?" Promise TV Web site, accessed March 23, 2013, www.promise.tv/what-is-it.html.

99 *"You can integrate your new ideas more easily":* Douglas C. Engelbart, "Augmenting Human Intellect: A Conceptual Framework" (Menlo Park, CA: Stanford Research Institute, 1962), accessed March 23, 2013, www.invisiblerevolution.net/engelbart/ full_62_paper_augm_hum_int.html.

99 *when we use word processors we're more iterative:* Christina Haas, "How the Writing Medium Shapes the Writing Process: Effects of Word Processing on Planning," *Research in the Teaching of English* 23, no. 2 (May 1989): 181–207; Ronald D. Owston, Sharon Murphy, and Herbert H. Wideman, "The Effects of Word Processing on Students' Writing Quality and Revision Strategies," *Research in the Teaching of English* 26, no. 3 (October 1992): 249–76; Merav Asaf and Ely Kozminsky, "Writing

Processes with Word Processors," *Proceedings of the Chais Conference on Instructional Technologies Research 2011: Learning in the Technological Era* (2011), accessed March 23, 2012, www.openu.ac.il/research_center/chais2011/download/writing_processes.pdf.

100 *a mesmerizing minute-long spectacle of memento mori:* clickflashwhirr (Web site), accessed March 23, 2013, clickflashwhirr.me/.

100 *There are "supercut" videos that combine dozens of common tropes:* naomileib, "Tasting Rachael Ray," YouTube, Oct 23, 2006, accessed March 23, www.youtube.com/watch?v=wRF_lewLjAM; Daniel Reis, "H@ckers ♥ Le@ther: A Style Supercut," YouTube, December 12, 2011, accessed March 23, 2013, www.youtube.com/watch?feature=player_embedded&v=pTkKwxsY2zE.

100 *YouTube creator Kutiman:* "Kutiman's Channel," YouTube, accessed March 23, 2013, www.youtube.com/user/kutiman.

100 *neatly critiqued the creepy, stalkerlike behavior of the lead vampire in* Twilight*:* Jonathan McIntosh, "Buffy vs Edward: Twilight Remixed—[original version]," YouTube, June 19, 2009, accessed March 23, 2013, www.youtube.com/watch?v=RZwM3GvaTRM.

101 *Consider an experiment conducted by MadV:* Some of my writing appeared previously in "Clive Thompson on How YouTube Changes the Way We Think," *Wired*, October 2011, accessed March 23, 2013, www.wired.com/techbiz/people/magazine/17-01/st_thompson.

102 *We're in the transitional moment:* Marshall McLuhan, *Understanding Media: The Extensions of Man*, critical ed. (1964; Corte Madera, CA: Gingko Press, 2003), 19–20; Eric Norden, "The Playboy Interview: Marshall McLuhan," *Playboy*, March 1969, reprinted in *Next Nature*, December 24, 2009, accessed March 23, 2013, www.nextnature.net/2009/12/the-playboy-interview-marshall-mcluhan/.

103 *"You can describe in words how to swing a golf club":* Clive Thompson, "The Know-It-All Machine," *Lingua Franca*, September 2001, accessed March 23, 2013, linguafranca.mirror.theinfo.org/print/0109/cover.html.

103 *the polish and dramatic flourish of presentations shot upward:* Chris Anderson, "TED Curator Chris Anderson on Crowd Accelerated Innovation," *Wired*, December 27, 2010, accessed March 23, 2013, www.wired.com/magazine/issue/19-01/.

104 *"Excuse these scraps of paper":* Lyman Horace Weeks, *A History of Paper-Manufacturing in the United States, 1690–1916* (New York: Lockwood Trade Journal Company, 1916), 46.

105 *Consider the case of Alexei Navalny:* Ellen Barry, "Rousing Russia with a Phrase," *The New York Times*, December 9, 2011, accessed March 23, 2013, www.nytimes.com/2011/12/10/world/europe/the-saturday-profile-blogger-aleksei-navalny-rouses-russia.html; Tom Parfitt, "Russian Protest Leader Alexei Navalny Is Target of Fake Photo Smear," *The Guardian*, January 9, 2012, accessed March 23, 2013, www.guardian.co.uk/world/2012/jan/09/russian-navalny-fake-photo-smear; Andrew E. Kramer, "Smear in Russia Backfires, and Online Tributes Roll In," *The New York Times*, January 8, 2012, accessed March 23, 2013, www.nytimes.com/2012/01/09/world/europe/smear-attempt-against-protest-leader-backfires-in-russia.html.

107 *online bloggers, schooled in Photoshop, smelled a rat:* Noah Shachtman, "Iran Missile Photo Faked (Updated)," *Danger Room* (*Wired* blog), July 10, 2008, accessed March 23, 2013, www.wired.com/dangerroom/2008/07/iran-missile-ph/.

108 *When Xeni Jardin of the* Boing Boing *blog:* Xeni Jardin, "Ralph Lauren Opens New Outlet Store in the Uncanny Valley," *Boing Boing*, September 29, 2009, accessed March 23, 2013, boingboing.net/2009/09/29/ralph-lauren-opens-n.html; Cory Doctorow, "The criticism that Ralph Lauren doesn't want you to see!" *Boing Boing*, October 6, 2009, accessed March 23, 2013, boingboing.net/2009/10/06/the -criticism-that-r.html; Mark Frauenfelder, "Searching for the Skinny on Ralph Lauren Ad (UPDATE: "We are Responsible," Says Ralph Lauren)," *Boing Boing*, October 8, 2009, accessed March 23, 2013, boingboing.net/2009/10/08/searching-for-the -sk.html.

109 *two trains collided near Wenzhou:* An Xiao Mina, "Social Media Street Art Responds to Chinese Train Disaster," *Hyperallergic*, August 4, 2011, accessed March 23, 2013, hyperallergic.com/31517/social-media-street-art/.

110 *And yet there's a deeper aspect to filtering:* Parts of my writing here on Instagram come from a column I wrote for *Wired* on the subject: "Clive Thompson on the Instagram Effect," *Wired*, January 2012, accessed March 23, 2013, www.wired .com/magazine/2011/12/st_thompson_instagram/.

110 *"seeing photographically":* Patrick Maynard, *The Engine of Visualization: Thinking through Photography* (Ithaca, NY: Cornell University Press, 2000), 197; Edward Weston, "Photography—Not Pictorial," *Camera Craft* 7, no. 37 (July 1930): 313– 20, accessed March 23, 2013, isites.harvard.edu/fs/docs/icb.topic639739.files/ Week%209/CB30-Weston-Photography.pdf.

110 *"The illiterate of the future":* Donis A. Dondis, *A Primer of Visual Literacy* (Cambridge, MA: MIT Press, 1974), xi.

111 *start-ups like MakerBot sell home 3-D printers starting at eighteen hundred dollars:* "The Replicator," MakerBot Web site, accessed March 23, 2013, store.makerbot .com/replicator.html.

Chapter 5: The Art of Finding

115 *cultures worldwide have a phrase for it:* Schacter, *The Seven Sins of Memory*, 72.

116 *"I can no longer function by myself":* Don Norman, "I Have Seen the Future and I Am Opposed," *Core77*, February 14, 2011, accessed March 23, 2103, www.core77.com/ blog/columns/i_have_seen_the_future_and_i_am_opposed_18532.asp.

116 *"When Wikipedia has a server outage":* Randall Munroe, "Extended Mind," *xkcd*, accessed March 24, 2013, http://www.xkcd.com/903/.

116 *Mesopotamian merchants began tallying their wares:* Alex Wright, *Glut: Mastering Information Through the Ages,* paperback ed. (Ithaca, NY: Cornell University Press, 2008), 48–50.

117 *The dialogue begins when Socrates encounters his young friend Phaedrus:* Plato, *Phaedrus*, Kindle edition.

117 *memorizing material was, in Socrates' day, a key cognitive skill for the elites:* Frances A. Yates, *The Art of Memory* (Chicago: University of Chicago Press, 1966), 1–48; Joshua Foer, *Moonwalking with Einstein* (New York: Penguin, 2011), 94–96.

118 *"As it was so easy for scholars to get books":* Blair, *Too Much to Know*, Kindle edition.

119 *"Memory is a rich and precious jewel":* Marius D'Assigny, *The Art of Memory*, 3rd ed. (1706), vii, accessed March 24, 2013, books.google.com/books?id=jy8CAAAAQ AAJ&printsec=frontcover#v=onepage&q&f=false.

119 *"Tumblers; Buffones & Juglers":* Wright, *Glut,* 131–32.

119 *"confusing the mind and disturbing clear ideas":* Blair, *Too Much to Know,* Kindle edition.

120 *"a man of no retentiveness":* Michel de Montaigne, *The Complete Works: Essays, Travel Journal, Letters,* trans. Donald M. Frame (New York: Everyman's Library, 2003), 359.

120 *he'd peruse a handful of unrelated books:* Blair, *Too Much to Know,* Kindle edition, and a personal interview with the author.

120 *Jeremias Drexel, a sixteenth-century Jesuit scholar of the humanities:* Ibid.

121 *A core job of the librarian . . . to conduct complex research:* Matthew Battles, *Library: An Unquiet History,* paperback edition (New York: Norton, 2004), 133–55; and an e-mail exchange with the author.

122 *truncated his first name to "Melvil" and last name to "Dui":* Wright, *Glut,* 174.

122 *In 1910, the French information-science pioneer Paul Otlet:* Ibid., 184–92.

123 *Bush was interested in how computers could improve human thought:* Vannevar Bush, "As We May Think," *The Atlantic Monthly,* July 1945, accessed March 24, 2013, www.theatlantic.com/magazine/archive/1945/07/as-we-may-think/303881/.

124 *that Harvard psychologist Daniel Wegner:* My discussion of transactive memory draws from several of Wegner's writings, all accessed March 24, 2013: Daniel Wegner, "Transactive Memory: A Contemporary Analysis of the Group Mind," in *Theories of Group Behavior,* eds. Brian Mullen and George R. Goethals (Berlin: Springer Verlag, 1987), 185–208, www.wjh.harvard.edu/~wegner/pdfs/Wegner%20Transactive%20Memory.pdf; Daniel Wegner, Ralph Erber, and Paula Raymond, "Transactive Memory in Close Relationships," *Journal of Personality and Social Psychology* 61, no. 6 (1991): 923–29, www.wjh.harvard.edu/~wegner/pdfs/Wegner,Erber,&Raymond1991.pdf; Daniel Wegner, Toni Giuliano, and Paula T. Hertel, "Cognitive Interdependence in Close Relationships," in *Compatible and Incompatible Relationships,* ed. William John Ickes (Berlin, Springer Verlag, 1985), 253–76, www.wjh.harvard.edu/~wegner/pdfs/Wegner,%20Giuliano,%20&%20Hertel%20(1985)%20Cognitive%20interdependence.pdf; and Daniel Wegner, "Don't Fear the Cybermind," *The New York Times,* August 4, 2012, accessed March 24, 2013, www.nytimes.com/2012/08/05/opinion/sunday/memory-and-the-cybermind.html. I also draw on a personal interview with Betsy Sparrow.

125 *owners opened their encyclopedias, on average, only once a year:* Shane Greenstein and Michelle Devereux, "The Crisis at Encyclopædia Britannica," case study (Kellogg School of Management, 2006), 3, accessed March 23, 2013, www.kellogg.northwestern.edu/faculty/greenstein/images/htm/research/cases/encyclopaediabritannica.pdf.

125 *One group of researchers studied older couples:* Celia B. Harris, Paul G. Keil, John Sutton, and Amanda J. Barnier, "Collaborative Remembering: When Can Remembering with Others Be Beneficial?" *Conference of the Australasian Society for Cognitive Science* (2009), accessed March 24, 2013, www.maccs.mq.edu.au/news/conferences/2009/ASCS2009/pdfs/Harris.pdf.

126 *assembling an AM/FM radio:* Richard L. Moreland and Larissa Myaskovsky, "Exploring the Performance Benefits of Group Training: Transactive Memory or Improved Communication?" *Organizational Behavior and Human Decision Processes* 82, no. 1 (May 2000): 117–33, accessed March 24, 2013, www.owlnet.rice

.edu/~antonvillado/courses/09a_psyc630001/Moreland%20&%20Myaskovsky%20(2000)%20OBHDP.pdf.

126 *researchers followed 209 undergraduates:* Marina Jackson and Richard L. Moreland, "Transactive Memory in the Classroom," *Small Group Research* 40, no. 5 (October 2009): 508–34.

126 *higher-scoring groups didn't just recall facts better:* Kyle Lewis, Donald Lange, Lynette Gillis, "Transactive Memory Systems, Learning, and Learning Transfer," *Organization Science* 16, no. 6 (November–December 2005): 581–98.

126 *"Quite simply, we seem to record as much outside our minds":* Wegner, "Transactive Memory: A Contemporary Analysis," 187.

126 *"Couples who are able to remember things transactively":* Wegner, Erber, and Raymond, "Transactive Memory in Close Relationships," 924.

127 *Sparrow ran an experiment:* Betsy Sparrow, Jenny Liu, Daniel M. Wegner, "Google Effects on Memory: Cognitive Consequences of Having Information at Our Fingertips," *Science* 333, no. 6043 (August 2011): 776–78; and a personal interview with the author.

128 *In 1990, psychologist Walter Kintsch documented this:* Gabriel A. Radvansky, "Situation Models in Memory: Texts and Stories," in Cohen and Conway, eds., *Memory in the Real World,* 229–31.

129 *the Wikipedia page on "Drone attacks in Pakistan":* "Drone attacks in Pakistan," Wikipedia, accessed March 24, 2013, en.wikipedia.org/wiki/Drone_attacks_in_Pakistan.

130 *40 percent of all queries are acts of* remembering: Jaime Teevan, Eytan Adar, Rosie Jones, and Michael A. S. Potts, "Information Re-Retrieval: Repeat Queries in Yahoo's Logs," in *SIGIR '07: Proceedings of the 30th Annual International ACM SIGIR Conference on Research and Development in Information Retrieval* (2007), 151–58.

131 *collaborative inhibition:* Celia B. Harris, Paul G. Keil, John Sutton, and Amanda J. Barnier, "We Remember, We Forget: Collaborative Remembering in Older Couples," *Discourse Processes* 48, no. 4 (2011), 267–303.

131 *In his essay "Mathematical Creation":* Henri Poincaré, "Mathematical Creation," in *The Anatomy of Memory: An Anthology* (New York: Oxford University Press, 1996), 126–35. I'm grateful to Jim Holt for drawing my attention to Poincaré's essay in Holt's essay "Smarter, Happier, More Productive," *London Review of Books* 33, no. 5 (March 3, 2011), 9–12, accessed March 24, 2013, www.lrb.co.uk/v33/n05/jim-holt/smarter-happier-more-productive; my analysis here draws on Holt's writing.

132 *a lament for the demise of memorization:* Clara Claiborne Park, "The Mother of the Muses: In Praise of Memory," in *Rejoining the Common Reader: Essays, 1962–1990* (Evanston, IL: Northwestern University Press, 1991), 20–39.

133 *a detailed description of a fictitious half inning of baseball:* George J. Spilich, Gregg T. Vesonder, Harry L. Chiesi, and James F. Voss, "Text Processing of Domain-Related Information for Individuals with High and Low Domain Knowledge," *Journal of Verbal Learning and Verbal Behavior* 18 (1979): 275–90.

133 *map experts retained far more details:* Kenneth J. Gilhooly, Michael Wood, Paul R. Kinnear, and Caroline Green, "Skill in Map Reading and Memory for Maps," *The Quarterly Journal of Experimental Psychology, Section A: Human Experimental Psychology* 40, no. 1 (1988): 87–107.

133 *memory for facts is quite specific to our obsessions:* This is a point Foer addresses eloquently in *Moonwalking,* 82–85 and 207–09; in the latter passage he also discusses the baseball study of Gilhooly et al.

135 *informatics professor Gloria Mark studied workplace employees:* Gloria Mark, Victor M. Gonzalez, and Justin Harris, "No Task Left Behind? Examining the Nature of Fragmented Work," *Proceedings of the SIGCHI Conference on Human Factors in Computing Systems* (2005): 321–30.

136 *Other research has confirmed that rapid task switching:* Eyal Ophir, Clifford Nass, and Anthony D. Wagner, "Cognitive Control in Media Multitaskers," *Proceedings of the National Academy of Sciences* 106, no. 37 (September 15, 2009): 15583–87, accessed March 24, 2013, www.pnas.org/content/106/37/15583.full.pdf+html, and personal interviews with Nass and Ophir; Amanda C. Gingerich, "Multi-Tasking = Epic Fail: Students Who Text Message During Class Show Impaired Comprehension of Lecture Material," poster, Midwest Institute for Students and Teachers of Psychology (2011), accessed March 23, 2013, digitalcommons.butler.edu/cgi/viewcontent.cgi?article=1200&context=facsch_papers.

136 *the "juggler's brain":* Carr, *Shallows,* 115.

136 *One survey of 197 businesses:* Vangelis Souitaris and B. M. Marcello Maestro, "Polychronicity in Top Management Teams: The Impact on Strategic Decision Processes and Performance of New Technology Ventures," *Strategic Management Journal* 31 (2010): 652–78, accessed March 24, 2013, www.cassknowledge.com/sites/default/files/article-attachments/413~~vangelissouitaris_polychronicity_in_top_management_teams.pdf.

136 *evidence suggests we comprehend as much:* Jordan T. Schugar, Heather Schugar, and Christian Penny, "A Nook or a Book? Comparing College Students' Reading Comprehension Levels, Critical Reading, and Study Skills," *International Journal of Technology in Teaching and Learning* 7, no. 2 (2011); 174–92; Jessica E. Moyer, " 'Teens Today Don't Read Books Anymore': A Study of Differences in Comprehension and Interest across Formats" (PhD dissertation, University of Minnesota, 2011), accessed March 24, 2013, readingformatchoicesdissertation.pbworks.com/f/ReadingFormatsFinalJuly27.pdf; Rebecca Dawn Baker, "Comparing the Readability of Text Displays on Paper, E-book Readers, and Small Screen Devices" (PhD dissertation, University of North Texas, 2010), accessed March 24, 2013, digital.library.unt.edu/ark:/67531/metadc28390/m2/1/high_res_d/dissertation.pdf. One limitation of all these studies is that in addition to being provisional, they measure comprehension, not retention. While the former is likely tied to the latter, it's not clear how.

137 *When Gloria Mark ran an experiment:* Gloria J. Mark, Stephen Voida, and Armand V. Cardello, " 'A Pace Not Dictated by Electrons': An Empirical Study of Work without Email," *Proceedings of the SIGCHI Conference on Human Factors in Computing Systems* (2012): 555–64. Leslie A. Perlow describes similarly positive effects in *Sleeping with Your Smartphone* (Boston: Harvard Business Press, 2012), in documenting how a group of consultants with the Boston Consulting Group agreed to stay off their devices for a set number of hours per day, which they called predictable time off, and which in saner times was referred to as evenings and weekends.

137 *"an overarching ability to watch and understand your own mind":* Maggie Jackson and Bill McKibben, *Distracted: The Erosion of Attention and the Coming Dark Age* (Amherst, NY: Prometheus Books, 2008), Kindle edition.

139 *The first wearable was cocreated in 1960 by Claude Shannon:* Edward O. Thorp, "The Invention of the First Wearable Computer," *Proceedings of the 2nd IEEE International Symposium on Wearable Computers* (1998): 4–8, accessed March 23, 2013, graphics.cs.columbia.edu/courses/mobwear/resources/thorp-iswc98.pdf.

141 *Critics have already noted how unsettling it might feel:* Mark Hurst, "The Google Glass Feature No One Is Talking About," *Creative Good* (blog), February 28, 2013, accessed March 24, 2013, creativegood.com/blog/the-google-glass-feature-no-one -is-talking-about/; Adrian Chen, "If You Wear Google's New Glasses You Are an Asshole," *Gawker*, March 3, 2013, accessed March 24, 2013, http://gawker .com/5990395.

144 *Their recall improved, sometimes dramatically:* Steve Hodges, Lyndsay Williams, Emma Berry, Shahram Izadi, James Srinivasan, Alex Butler, Gavin Smyth, Narinder Kapur, and Ken Wood, "SenseCam: A Retrospective Memory Aid," *Proceedings of the 8th International Conference of Ubiquitous Computing* (2006): 177–93; Georgina Browne, Emma Berry, Narinder Kapur, Steve Hodges, Gavin Smyth, Peter Watson, and Ken Wood, "SenseCam Improves Memory for Recent Events and Quality of Life in a Patient with Memory Retrieval Difficulties," *Memory* 19, no. 7 (2011): 713–22; Georgina Browne, Emma Berry, Steve Hodges, Gavin Smyth, Alex Butler, Lyndsay Williams, James Srinivasan, Alban Rrustemi, and Ken Wood, "Stimulating Episodic Memory Using SenseCam," poster presentation on Microsoft Research Web site (2007), accessed March 24, 2013, research.microsoft.com/pubs/132686/4%20 Festival%20of%20Internation%20Conferences%20Poster%20.pdf; and personal interview with Lyndsay Williams and Ken Wood.

144 *this process could work in reverse:* Ebbinghaus, *Memory*, Kindle edition.

144 *"spaced repetition":* John J. Donovan and David J. Radosevich, "A Meta-Analytic Review of the Distribution of Practice Effect: Now You See It, Now You Don't," *Journal of Applied Psychology* 84, no. 5 (October 1999): 795–805.

145 *in an Ebbinghausian fashion:* "Frequently Asked Questions," Amazon Kindle Web site, accessed March 24, 2013, kindle.amazon.com/faq.

Chapter 6: The Puzzle-Hungry World

147 *"I've poured over a hundred hours into the game":* Paul Tassi, "Why Skyrim Is Not My Game of the Year," *Forbes*, December 22, 2011, accessed March 24, 2103, www .forbes.com/sites/insertcoin/2011/12/22/why-skyrim-is-not-my-game-of-the-year/.

147 *only a few lines of instruction:* Harold Goldberg, "The Origins of the First Arcade Video Game: Atari's *Pong*," *Vanity Fair*, March 28, 2011, accessed March 24, 2013, www.vanityfair.com/culture/features/2011/03/pong-excerpt-201103.

148 *the "no fire" bug:* Christopher Cantrell, "The Galaga No-Fire Cheat," Computer Archeology, accessed March 24, 2013, computerarcheology.com/galaga/galaga.html.

149 *programmed into the Atari game* Adventure: "The Greatest Easter Eggs in Gaming," Gamespot, accessed March 24, 2013, www.gamespot.com/features/the-greatest -easter-eggs-in-gaming-6131572/; Joey Connelly, "Of Dragons and Easter Eggs: A Chat with Warren Robinett," *The Jaded Gamer*, May 13, 2003, accessed March 24, 2013, thejadedgamer.com/of-dragons-and-easter-eggs-a-chat-with-warren -robinett/. Technically the *Adventure* Easter egg preceded the industry's appreciation for the networked intelligence of gamers, though it gave birth to the phrase "Easter

egg," when, as Robinett notes in the latter piece here, they told him "it's kind of cool to have little hidden surprises in video games. It's kinda like waking up on Easter morning and hunting for Easter eggs."

149 *"double punch–elbow–double-arm suplex throw":* Stephen Hamilton, *Virtua Fighter 3 FAQs*, last revised March 5, 1997, accessed March 24, 2013, www.gamefaqs.com/arcade/583649-virtua-fighter-3/faqs/1127.

150 *a wiki with 15,789 pages: "The Elder Scrolls V: Skyrim,"* The *Elder Scrolls* Wiki, elderscrolls.wikia.com/wiki/The_Elder_Scrolls_V:_Skyrim. When I checked this wiki a few months after the game launched, it had 15,789 pages; when last accessed on March 23, 2013, it had 18,776.

150 *"You can get an invincible companion dog":* Skkragggh, "Ye Grande Olde Liste of Skyrim Tips n' Tricks," The *Elder Scrolls* Wiki, last updated April 9, 2012, accessed March 24, 2013, http://elderscrolls.wikia.com/wiki/User_blog:Skkragggh/Ye_Grande_Olde_Liste_of_Skyrim_Tips_n'_Tricks.

150 *most players use crowdsourced documents sparingly:* Mia Consalvo, *Cheating: Gaining Advantage in Videogames* (Cambridge, MA: MIT Press, 2009); and a personal interview with the author. This writing draws on previous reporting I've done on Consalvo's work, specifically "Clive Thompson on Puzzles and the Hive Mind," *Wired*, April 2009, accessed March 24, 2013, www.wired.com/techbiz/people/magazine/17-05/st_thompson, and "What Type of Game Cheater Are You?" *Wired*, April 23, 2007, accessed March 24, 2013, www.wired.com/gaming/virtualworlds/commentary/games/2007/04/gamesfrontiers_0423.

150 *According to one estimate in 2008:* Barb Dybwad, "SXSW08: How Gamers Are Adopting the Wiki Way," Massively by Joystiq, March 8, 2008, accessed March 24, 2013, massively.joystiq.com/2008/03/08/sxsw08-how-gamers-are-adopting-the-wiki-way/.

151 *society has always had latent groups . . . nothing got done:* Clay Shirky, *Here Comes Everybody: The Power of Organizing without Organizations* (New York: Penguin, 2008), Kindle edition.

152 *"under the Coasean floor":* Ibid.

154 *the publication of* Fifty Shades of Grey: Julie Bosman, "Discreetly Digital, Erotic Novel Sets American Women Abuzz," *The New York Times*, March 9, 2012, accessed March 23, 2013, www.nytimes.com/2012/03/10/business/media/an-erotic-novel-50-shades-of-grey-goes-viral-with-women.html?pagewanted=all&_r=0.

154 *"If you wanted to read a 3000 word fic":* Maciej Cegłowski, "The Fans Are All Right," *Pinboard Blog*, accessed March 24, 2013, blog.pinboard.in/2011/10/the_fans_are_all_right/.

155 *"These people," he wrote later, "do not waste time":* Ibid.

156 *"the crowd's judgment was essentially perfect":* James Surowiecki, *The Wisdom of Crowds: Why the Many Are Smarter Than the Few and How Collective Wisdom Shapes Business, Economies, Societies, and Nations* (New York: Doubleday, 2004), xiii.

159 *the spectacular collapse of the* Los Angeles Times's *"wikitorial":* Dan Glaister, "LA Times 'wikitorial' gives editors red faces," *The Guardian*, June 21, 2005, accessed March 25, 2013, www.guardian.co.uk/technology/2005/jun/22/media.pressandpublishing.

159 *an online experiment to group-think pictures:* Kevan Davis, "The Smaller Picture" (2002), accessed March 24, 2013, kevan.org/smaller.cgi; I previously wrote about

the project in "Art Mobs," *Slate*, July 21, 2004, accessed March 24, 2013, www
.slate.com/articles/technology/webhead/2004/07/art_mobs.html.

160 *"The small contributions help the collaboration":* Michael Nielsen, *Reinventing Discovery: The New Era of Networked Science* (Princeton, NJ: Princeton University
Press, 2011), 64.

161 *there are not quite thirty-five hundred:* "Wikipedia Statistics English," Wikipedia,
stats.wikimedia.org/EN/TablesWikipediaEN.htm. When I last accessed this page on
March 24, 2013, it showed that the number of contributors who made more than one
hundred edits in January 2013 was 3,414.

161 *the* really *committed folks—the administrators:* "Wikipedia:List of administrators,"
Wikipedia, en.wikipedia.org/wiki/Wikipedia:List_of_administrators. When I last
accessed this page on March 24, 2013, it listed the number of active administrators
as 690. The role of administrators in dealing with vandalism is noted in Joseph Michael Reagle Jr., *Good Faith Collaboration: The Culture of Wikipedia* (Cambridge,
MA: MIT Press, 2010), Kindle edition.

162 *"Kasparov versus the World":* Nielsen, *Reinventing Discovery*, 15–18.

162 *"a reasonably serious but not elite junior player":* Ibid., 26.

162 *"the greatest game in the history of chess":* Michael Nielsen, "Kasparov versus the
World," *Michael Nielsen* (blog), August 21, 2007, accessed March 24, 2013, mi
chaelnielsen.org/blog/kasparov-versus-the-world/.

163 *"Mutual respect and a reasonable approach to disagreement":* Jimbo Wales, "Letter
from the Founder," Wikimedia Foundation, September 2004, accessed March 24,
2013, wikimediafoundation.org/wiki/Founder_letter/Founder_letter_Sept_2004.

163 *Wikipedia's Five Pillars of self-governance:* "Wikipedia:Five pillars," accessed March
24, 2103, en.wikipedia.org/wiki/Wikipedia:Five_pillars.

164 *"there is plenty of time to stop and ask questions . . . they need to be cordial on Wikipedia":* Reagle, *Good Faith Collaboration*, Kindle edition.

164 *social scientists tested different techniques of brainstorming:* Frans Johansson, *The
Medici Effect: What Elephants and Epidemics Can Teach Us about Innovation*
(Boston: Harvard Business School Press, 2006), 108–10; and Susan Cain, *Quiet:
The Power of Introverts in a World That Can't Stop Talking* (New York: Crown,
2012), Kindle edition. My writing here also draws on a column I previously wrote on
this subject: "Clive Thompson on the Power of Introversion," *Wired*, April 2012,
accessed March 24, 2013, www.wired.com/magazine/2012/03/st_thompson_
introvert/.

165 *A 2011 study took several virtual groups:* Jan Lorenza, Heiko Rauhut, Frank
Schweitzera, and Dirk Helbing, "How Social Influence Can Undermine the Wisdom
of Crowd Effect," *PNAS* 108, no. 22 (May 31, 2011): 9020–25, accessed March 24,
2013, www.pnas.org/content/108/22/9020.full.pdf#page=1&view=FitH. The "rich
get richer" problem has been investigated extensively by the network scientist Duncan J. Watts; he writes about his research in "Is Justin Timberlake a Product of Cumulative Advantage?" *The New York Times Magazine*, April 15, 2007, accessed
March 24, 2013, www.nytimes.com/2007/04/15/magazine/15wwlnidealab.t.html,
and *Everything Is Obvious*: *Once You Know the Answer* (New York: Crown
Business, 2011): 54–81.

167 *"Encouraging discussion and questions":* "Contenders," Fold.it Web site, accessed
March 24, 2013, fold.it/portal/node/985857.

168 *Baker published the results:* Firas Khatib et al., "Crystal Structure of a Monomeric Retroviral Protease Solved by Protein Folding Game Players," *Nature Structural & Molecular Biology* 19, no. 3 (March 2012): 1175–77.

169 *the Galaxy Zoo:* Tim Adams, "Galaxy Zoo and the New Dawn of Citizen Science," *The Observer* (UK), March 17, 2012, accessed March 24, 2013, www.guardian .co.uk/science/2012/mar/18/galaxy-zoo-crowdsourcing-citizen-scientists.

169 *a one-million-dollar prize:* Eliot Van Buskirk, "BellKor's Pragmatic Chaos Wins $1 Million Netflix Prize by Mere Minutes," *Wired*, September 21, 2009, accessed March 24, 2013, www.wired.com/business/2009/09/bellkors-pragmatic-chaos-wins-1-million-netflix-prize/; I also previously reported on the Netflix Prize in "If You Liked This, You're Sure to Love That," *The New York Times Magazine*, November 21, 2008, accessed March 24, 2013, www.nytimes.com/2008/11/23/magazine/23Netflix-t.html. I didn't specifically note the increasing secrecy of the participants over time in my article, but the teams remarked on this in my interviews.

170 *an employee pitched a new product:* Jim Lavoie, "Nobody's as Smart as Everybody—Unleashing Individual Brilliance and Aligning Collective Genius," Management Innovation eXchange Web site, April 8, 2010, accessed March 24, 2013, www .managementexchange.com/story/nobody's-smart-everybody-unleashing-quiet -genius-inside-organization.

170 *Bill Joy reportedly liked to say:* David Bornstein, "Social Change's Age of Enlighten-ment," *Opinionator* (*New York Times* blog), October 17, 2012, accessed March 24, 2013, opinionator.blogs.nytimes.com/2012/10/17/social-changes-age-of -enlightenment/.

170 *a trove of expense receipts for British politicians:* My writing here derives from a previ-ous column on the subject, "Clive Thompson on How Games Make Work Seem Like Play," *Wired*, March 2011, accessed March 24, 2013, www.wired.com/maga zine/2011/02/st_thompson_living_games/. One detail has changed since then; the column reported a pen costing $441 (the U.S. equivalent of £225 when the column was written), but this receipt wasn't refindable when writing this book, whereas one for a £225 duvet was.

170 *fans of the Brian Lehrer radio show:* "Are You Being Gouged?" WNYC Web site, September 24, 2007, accessed March 24, 2013, www.wnyc.org/shows/bl/2007/ sep/24/are-you-being-gouged/.

172 *Humans are not ants:* Pierre Lévy, *Collective Intelligence: Mankind's Emerging World in Cyberspace*, trans. Robert Bononno (Cambridge, MA: Helix Books, 1999), 16–17.

172 *a better metaphor for collaborative thinking:* Sherlock Holmes: The quotes here are from the following works by Arthur Conan Doyle: the novel *The Sign of the Four* (Allan Classics, 2010), Kindle edition; and the story "The Adventure of Wisteria Lodge," Project Gutenberg, last updated December 2011, accessed March 24, 2013, www.gutenberg.org/files/2343/2343-h/2343-h.htm.

Chapter 7: Digital School

175 *When I visit Matthew Carpenter's math class:* Some of this reporting appeared origi-nally in an article I wrote about the Khan Academy, "How Khan Academy Is Chang-ing the Rules of Education," *Wired*, August 2011, accessed March 24, 2013, www .wired.com/magazine/2011/07/ff_khan/.

177 *the "Two Sigma" phenomenon:* Benjamin S. Bloom, "The 2 Sigma Problem: The Search for Methods of Group Instruction as Effective as One-to-One Tutoring," *Educational Researcher* 13, no. 6 (June–July 1984), 4–16; and Benjamin S. Bloom, "The Search for Methods of Group Instruction as Effective as One-to-One Tutoring," *Educational Leadership* 41, no. 8 (May 1984) 4–17.

177 *roughly twenty-five children:* Sarah D. Sparks, "Class-Size Limits Targeted for Cuts," *Education Week* 30, no. 13 (December 1, 2010): 1, 16, updated March 23, 2012, accessed March 25, 2013, www.edweek.org/ew/articles/2010/11/24/13size_ep.h30 .html.

178 *One U.S. federal study:* Barbara Means et al., "Evaluation of Evidence-Based Practices in Online Learning," U.S. Department of Education, revised September 2010, accessed March 24, 2013, www2.ed.gov/rschstat/eval/tech/evidence-based-practices/ finalreport.pdf.

178 *Laurentius de Voltolina painted a scene:* Nancy L. Turner, "Medieval University Classroom," accessed March 24, 2013, www.uwplatt.edu/~turnern/classroomFull.html.

179 *As various educational analysts have joked:* I've heard this joke from many educational experts, though one of the first I've heard it attributed to is Seymour Papert, in Joe Bower, "Time-Traveling Teachers," *For the Love of Learning* (blog), July 16, 2012, accessed March 24, 2013, www.joebower.org/2012/07/time-traveling-teachers.html.

180 *a long history of hype that has rarely delivered:* Ellen A. Wartella and Nancy Jennings, "Children and Computers: New Technology—Old Concerns," *The Future of Children* 10, no. 2 (Autumn–Winter, 2000): 31–43, accessed March 24, 2013, www .princeton.edu/futureofchildren/publications/docs/10_02_01.pdf; Bruce Lenthall, "Critical Reception: Public Intellectuals Decry Depression-Era Radio, Mass Culture, and Modern America," in *Radio Reader: Essays in the Cultural History of Radio*, eds. Michele Hilmes and Jason Loviglio (New York: Routledge, 2002), 41–62; Michael Haworth and Stephanie Hopkins, "On the Air: Educational Radio, Its History and Effect on Literacy and Educational Technology," *ETEC540 Community Weblog*, accessed March 24, 2013, blogs.ubc.ca/etec540sept09/2009/10/28/ on-the-air-educational-radio-its-history-and-effect-on-literacy-and-educational -technology-by-michael-haworth-stephanie-hopkins/.

180 *computers have been the most successful in infiltrating classrooms:* The sixty-billion-dollar estimate comes from Clayton M. Christensen, Michael B. Horn, and Curtis W. Johnson, *Disrupting Class: How Disruptive Innovation Will Change the Way the World Learns* (New York: McGraw-Hill, 2008), 11–12; Matt Richtel's story on Apple is "Silicon Valley Wows Educators, and Woos Them," *The New York Times*, November 4, 2011, accessed March 24, 2013, www.nytimes.com/2011/11/05/ technology/apple-woos-educators-with-trips-to-silicon-valley.html; the ratio of computers is in Christensen et al., *Disrupting Class*, 2–3.

183 *National Assessment of Educational Progress:* The NAEP studies mentioned here are: "Average Mathematics Scale Score, by Age and Selected Student and School Characteristics: Selected Years, 1973 through 2008," accessed March 24, 2013, nces.ed .gov/programs/digest/d11/tables/dt11_141.asp; "Average Reading Scale Score, by Age and Selected Student and School Characteristics: Selected Years, 1971 through 2008," accessed March 24, 2013, nces.ed.gov/programs/digest/d11/tables/dt11_125 .asp; Lawrence C. Stedman, "The NAEP Long-Term Trend Assessment: A Review of Its Transformation, Use, and Findings," National Assessment Governing Board Web

site, accessed March 24, 2013, www.nagb.org/content/nagb/assets/documents/who-we-are/20-anniversary/stedman-long-term-formatted.pdf.

184 *literacy scholar Steve Graham:* Steve Graham and Michael Hebert, *Writing to Read: Evidence for How Writing Can Improve Reading* (New York: Carnegie Corporation, 2010), 14, accessed March 24, 2013, carnegie.org/fileadmin/Media/Publications/WritingToRead_01.pdf

184 *Dorothy Burt, a literacy project facilitator in Point England:* All the material about the New Zealand school comes from personal interviews with Dorothy Burt, Colleen Gleeson, and Russell Burt.

187 *guide to a* Star Trek *game:* Eric Davey, "Star Trek Tactical Assault," GameFAQs, last updated August 22, 2008, accessed March 24, 2013, www.gamefaqs.com/ds/931492-star-trek-tactical-assault/faqs/49918.

187 *surveyed about their writing:* Amanda Lenhart, Sousan Arafeh, Aaron Smith, and Alexandra MacGill, *Writing, Technology and Teens* (Pew Internet & American Life Project, April 24, 2008), accessed March 24, 2013, www.pewinternet.org/Reports/2008/Writing-Technology-and-Teens/09-The-Way-Teens-See-Their-Writing.aspx?view=all.

188 *This was the epiphany of Seymour Papert:* My description of Papert's work and thinking draws from his book *Mindstorms: Children, Computers, and Powerful Ideas,* 2nd ed. (New York: Basic Books, 1993).

194 *download someone else's Scratch game to reverse engineer it:* Benjamin Mako Hill, Andrés Monroy-Hernández, and Kristina R. Olson, "Responses to Remixing on a Social Media Sharing Website," presented at the Fourth International Association for the Advancement of Artificial Intelligence Conference on Weblogs and Social Media (2010), accessed March 24, 2013, www.aaai.org/ocs/index.php/ICWSM/ICWSM10/paper/view/1533/1836.

194 *formed their own international groups:* Cecilia R. Aragon, Sarah S. Poon, and Andrés Monroy-Hernández, "A Tale of Two Online Communities: Fostering Collaboration and Creativity in Scientists and Children," *Proceedings of the Seventh ACM Conference on Creativity and Cognition* (2009): 9–18.

195 *"Learning from others":* Douglas Thomas and John Seely Brown, *A New Culture of Learning: Cultivating the Imagination for a World of Constant Change* (CreateSpace self-published book, 2011), 51.

195 *A 2007 U.S. government survey of 737 fifth-grade classrooms:* Robert C. Pianta, Jay Belsky, Renate Houts, and Fred Morrison, "Opportunities to Learn in America's Elementary Classrooms," *Science* 315, no. 5820 (March 30, 2007): 1795–96, accessed March 24, 2013, www.ncbi.nlm.nih.gov/pmc/articles/PMC2137172/.

195 *as media theorist Douglas Rushkoff argues:* Douglas Rushkoff, *Program or Be Programmed: Ten Commandments for a Digital Age* (New York: OR Books, 2010), 120, 133, 140.

196 *Steinkuehler had joined a guild:* Some of my writing here appeared previously in Clive Thompson, "How Videogames Blind Us with Science," *Wired,* September 8, 2008, accessed March 24, 2013, www.wired.com/gaming/gamingreviews/commentary/games/2008/09/gamesfrontiers_0908.

197 *involved "scientific" activity:* Constance Steinkuehler and Sean Duncan, "Scientific Habits of Mind in Virtual Worlds," *Journal of Science Education and Technology* 17, no. 6 (Dec 2008): 530–43.

198 *"algebra talk":* James Paul Gee, *Getting Over the Slump: Innovation Strategies to Promote Children's Learning* (New York: The Joan Ganz Cooney Center, 2008), 23.

198 *an ideal native environment for teaching the power of scientific rigor:* Constance Steinkuehler and Marjee Chmiel, "Fostering Scientific Habits of Mind in the Context of Online Play," *Proceedings of the 7th International Conference on Learning Sciences* (2006), 723–29, accessed March 24, 2013, website.education.wisc.edu/steinkuehlcr/blog/papers/tenure/publications/51_SteinkuehlerChmiel.pdf.

200 *invited to oversee a group of struggling students in a Boston high school:* My reporting on the *Civilization 3* experiment draws from Kurt Squire, *Video Games and Learning: Teaching and Participatory Culture in the Digital Age* (New York: Teachers College Press, 2011), 112–39, and a personal interview with the author.

202 *one physics teacher uses* Angry Birds*:* John Burk, "Why You Should Wait to Teach Projectile Motion Part 2: Introducing Projectile Motion Using Angry Birds," *Quantum Progress* (blog), February 17, 2011, accessed March 24, 2013, quantumprogress.wordpress.com/2011/02/17/why-you-should-wait-to-teach-projectile-motion-part-2-introducing-projectile-motion-using-angry-birds/.

202 *created* Supercharged!: Squire, *Video Games and Learning*, 90–100.

203 *purpose-driven "efferent" reading:* Louise M. Rosenblatt, *The Reader, the Text, the Poem*, paperback edition (Carbondale, IL: Southern Illinois University Press, 1994), 23–25.

203 *She took a group of boys:* Constance Steinkuehler, "The Mismeasure of Boys: Reading and Online Videogames" (WCER Working Paper No. 2011-3), accessed March 25, 2013, www.wcer.wisc.edu/publications/workingPapers/Working_Paper_No_2011_03.pdf.

204 *term "digital natives":* Marc Prensky, "Digital Natives, Digital Immigrants," originally in *On the Horizon* 9, no. 5 (October 2001), accessed March 25, 2013, www.marcprensky.com/writing/Prensky%20-%20Digital%20Natives,%20Digital%20Immigrants%20-%20Part1.pdf.

204 *test students' facility on an omnipresent digital skill:* Bing Pan et al., "In Google We Trust: Users' Decisions on Rank, Position, and Relevance," *Journal of Computer-Mediated Communication* 12, no. 3, article 3 (2007), accessed March 25, 2013, jcmc.indiana.edu/vol12/issue3/pan.html. Portions of my writing here and later in this chapter appeared previously in "Clive Thompson on Why Kids Can't Search," *Wired*, November 2011, accessed March 25, 2013, www.wired.com/magazine/2011/11/st_thompson_searchresults/.

204 *similarly dismal results:* Eszter Hargittai, Lindsay Fullerton, Ericka Menchen-Trevino, and Kristin Yates Thomas, "Trust Online: Young Adults' Evaluation of Web Content," *International Journal of Communication* 4 (2010), 468–94, accessed March 24, 2013, www.webuse.org/pdf/HargittaiEtAlTrustOnlineIJoC10.pdf; Eszter Hargittai, "The Role of Expertise in Navigating Links of Influence," in *The Hyperlinked Society: Questioning Connections in the Digital Age*, eds. Joseph Turow and Lokman Tsui (Ann Arbor, MI: University of Michigan Press, 2008), 85–103.

206 *Crap detection:* Howard Rheingold, "Crap Detection 101," *SFGate*, June 30, 2009, accessed March 25, 2013, blog.sfgate.com/rheingold/2009/06/30/crap-detection-101/. As Rheingold notes, he's modifying the phrase from Ernest Hemingway's original witticism: "Every man should have a built-in automatic crap detector operating inside him."

208 *"inept and satisfied end user":* T. Scott Plutchak, "Inept and Satisfied, Redux," *Journal of the Medical Library Association* 93, no. 1 (January 2005): 1–3, accessed online March 26, 2013, www.ncbi.nlm.nih.gov/pmc/articles/PMC545111/.

Chapter 8: Ambient Awareness

209 *Ben Haley, a technical support specialist in Seattle:* Portions of my writing in this chapter, including my reporting on Haley, appeared originally in "Brave New World of Digital Intimacy," *The New York Times Magazine*, September 5, 2008, accessed March 26, 2013, www.nytimes.com/2008/09/07/magazine/07awareness-t.html.

210 *a phrase for this type of ESP:* In addition to "ambient awareness," another evocative term for this phenomenon was coined by the user-experience expert Leisa Reichelt: "ambient intimacy." She wrote about it on her blog post "Ambient Intimacy," *Disambiguity*, March 1, 2007, accessed March 26, 2013, www.disambiguity.com/ambient -intimacy/.

210 *Mizuko Ito, a cultural anthropologist, first noticed this effect:* Mizuko Ito and Daisuke Okabe, "Technosocial Situations: Emergent Structuring of Mobile E-mail Use," in *Personal, Portable, Pedestrian: Mobile Phones in Japanese Life*, eds. Mizuko Ito, Misa Matsuda, and Daisuke Okabe (Cambridge, MA: MIT Press, 2005), 261–73.

211 *"It's like you're in the room":* Some of my writing here appeared previously in Clive Thompson, "Remote Possibilities," *The New York Times Magazine*, November 16, 2003, accessed March 26, 2013, www.nytimes.com/2003/11/16/magazine/16CELL .html.

213 *studying the coordination of staff in the control rooms:* This material comes from Christian Heath and Paul Luff, "Collaboration and Control: Crisis Management and Multimedia Technology in London Underground Line Control Rooms," *Journal of Computer Supported Cooperative Work* 1, no. 1 (1992): 24–48; Lucy Suchman, "Centers of Coordination: A Case and Some Themes," in *Discourse, Tools, and Reasoning: Essays on Situated Cognition*, eds. Lauren B. Resnick, Roger Säljö, Clotilde Pontecorvo, and Barbara Burge (Berlin: Springer Verlag, 1997), 41–62, accessed March 26, 2013, www.ida.liu.se/~729G12/mtrl/Suchman_Centres_of_coordina tion.pdf; and a personal interview with Heath.

213 *a form of proprioception:* Portions of my writing here appeared in "Clive Thompson on How Twitter Creates a Social Sixth Sense," *Wired*, June 2007, accessed March 26, 2013, www.wired.com/techbiz/media/magazine/15-07/st_thompson.

216 *have strangers inspect the Facebook pages of experimental subjects:* My description of Gosling's work comes from Samuel D. Gosling, Sei Jin Ko, Thomas Mannarelli, and Margaret E. Morris, "A Room with a Cue: Personality Judgments Based on Offices and Bedrooms," *Journal of Personality and Social Psychology* 82, no. 3 (2002): 379–98, accessed March 26, 2013, homepage.psy.utexas.edu/homepage/faculty/Gosling/ reprints/JPSP02-Roomwithacue.pdf; Sam Gosling, *Snoop: What Your Stuff Says About You* (New York: Basic Books, 2009), Kindle edition; Samuel D. Gosling, Sam Gaddis, and Simine Vazire, "Personality Impressions Based on Facebook Profiles," *Proceedings of ICWSM 2007*, accessed March 26, 2013, www.icwsm.org/papers/3 —Gosling-Gaddis-Vazire.pdf; and a personal interview with Gosling.

216 *the work patterns of younger employees in the United States:* My writing here draws on a personal interview with Michael O'Neill and his report *Generational Preferences:*

A Glimpse into the Future Office (Knoll Inc., 2010), accessed March 26, 2013, www.knoll.com/research/downloads/WP_GenerationalDifferences.pdf. Some of this work appeared previously in Clive Thompson, "The End of the Office," *The Globe and Mail* (Toronto), June 17, 2010, accessed March 26, 2013, www.the globeandmail.com/report-on-business/small-business/the-end-of-the-office/article 4322237/?page=all.

217 *the phone call:* This writing draws on a previous column of mine, "Clive Thompson on the Death of the Phone Call," *Wired*, August 2010, accessed March 26, 2013, www.wired.com/magazine/2010/07/st_thompson_deadphone/.

218 *Office workers spend an estimated 28 percent of the workweek:* Michael Chui et al., *The Social Economy: Unlocking Value and Productivity through Social Technologies* (McKinsey Global Institute, 2012), 46, accessed March 26, 2013, www.mckin sey.com/insights/high_tech_telecoms_internet/the_social_economy; "Email Statistics Report, 2011–2015," The Radicati Group, May 2011, accessed March 26, 2013, www.radicati.com/wp/wp-content/uploads/2011/05/Email-Statistics-Report-2011 -2015-Executive-Summary.pdf.

219 *a survey of 912 white-collar workers:* R. Kelly Garrett and James N. Danziger, "IM = Interruption Management? Instant Messaging and Disruption in the Workplace," *Journal of Computer-Mediated Communication* 13, no. 1 (2007): 23–42, accessed March 26, 2013, jcmc.indiana.edu/vol13/issue1/garrett.html.

220 *companies are trying to follow his lead:* "Atos Boss Thierry Breton Defends His Internal Email Ban," BBC News, December 6, 2011, accessed March 26, 2013, www.bbc .co.uk/news/technology-16055310; "Volkswagen Turns Off Blackberry Email after Work Hours," BBC News, December 23, 2011, accessed March 26, 2013, www.bbc .co.uk/news/technology-16314901.

220 *A study by the McKinsey Global Institute:* Chui et al., *The Social Economy*, 11.

220 *"Who really cares what I am doing":* Alex Beam, "Twittering with Excitement? Hardly," *Boston Globe*, E4, August 16, 2008.

221 *Twitter, as the editors of the literary magazine* n+1 *despaired:* "Please RT," *n+1* 14 (Summer 2012), accessed March 26, 2013, nplusonemag.com/please-rt.

221 *"the enemy of contemplation":* Bill Keller, "The Twitter Trap," *The New York Times Magazine*, May 18, 2011, accessed March 26, 2013, www.nytimes.com/2011/05/22/ magazine/the-twitter-trap.html.

221 *Social critics, including Christopher Lasch:* Christopher Lasch, *The Culture of Narcissism: American Life in an Age of Diminishing Expectations* (New York: Norton, 1991).

221 *empirical proof of an epidemic of narcissism:* Jean Twenge summarizes her research finding that narcissism is rising in her book written with W. Keith Campbell, *The Narcissism Epidemic: Living in the Age of Entitlement* (New York: Simon and Schuster, 2010); more recently, she has documented a decades-long increase in how often American book authors use words that reflect "individualistic traits," in Jean M. Twenge, W. Keith Campbell, and Brittany Gentile, "Changes in Pronoun Use in American Books and the Rise of Individualism, 1960–2008," *Journal of Cross-Cultural Psychology* 44, no. 3 (April 2013), 406–15. However, the work of Kali H. Trzesniewski has critiqued Twenge's findings on several fronts, arguing that psychological tools for measuring narcissism are not necessarily reliable and that other data show young people's personalities have remained quite stable over several decades: Kali H. Trzesniewski,

M. Brent Donnellan, and Richard W. Robins, "Is 'Generation Me' Really More Narcissistic Than Previous Generations?" *Journal of Personality* 76, no. 4 (August 2008):
903–18; Kali H. Trzesniewski and M. Brent Donnellan, "Rethinking 'Generation Me':
A Study of Cohort Effects From 1976–2006," *Perspectives on Psychological Science* 5,
no. 1 (2010): 58–75. A good layperson's guide to this debate is Sadie F. Dingfelder,
"Reflecting on Narcissism," *Monitor on Psychology* 42, no. 2 (February 2011), accessed March 26, 2013, www.apa.org/monitor/2011/02/narcissism.aspx.

222 *an inevitable side effect of mass publishing:* Clay Shirky, *Cognitive Surplus: Creativity and Generosity in a Connected Age* (New York: Penguin, 2010), 47.

222 *Robin Dunbar studied everyday face-to-face conversation:* Robin Dunbar, *Grooming,
Gossip, and the Evolution of Language*, paperback ed. (Cambridge, MA: Harvard
University Press, 1998), 123, 176, 192–93; R. I. M. Dunbar, Anna Marriott, and
W. D. C. Duncan, "Human Conversational Behavior," *Human Nature* 8, no.
3(1997): 238.

223 *As Tom Standage notes:* Tom Standage, "The Distractions of Social Media, 1673
Style," Tomstandage.com, June 12, 2012, accessed March 26, 2013, https://tom
standage.wordpress.com/2012/06/12/the-distractions-of-social-media-1673-style/.

223 *the 1835 essay "Devouring Books":* "Devouring Books," in *American Annals of Education and Instruction for the Year 1835*, ed. William C. Woodbridge (Boston: Light
& Horton, 1835), 30–32, accessed via Google Books, March 26, 2013, books
.google.com/books?id=8q8pAQAAMAAJ.

224 *As Jonathan Swift put it:* "On the Education of Ladies," in *The Works of Dr. Jonathan Swift, Dean of St. Patrick's, Dublin, Volume XVII* (London: W. Johnston,
1766), 71, accessed via Google Books, March 26, 2013, books.google.com/
books?id=91FY-OOnWi0C.

224 *The U.S. neurologist George Miller Beard:* George M. Beard, *American Nervousness,
Its Causes and Consequences* (New York: G. P. Putnam's Sons, 1881), vi, accessed
via Google Books, March 26, 2013, books.google.com/books?id=3moPAAAAYAAJ.

224 *Who would bother to leave the house:* Claude S. Fisher, *America Calling: Social History of the Telephone to 1940* (Berkeley: University of California Press, 1992), 224.

224 *"A Telephonic Conversation":* Mark Twain, "A Telephonic Conversation," *The Atlantic Monthly*, June 1, 1880, accessed March 26, 2013, www.theatlantic.com/maga
zine/archive/1880/06/a-telephonic-conversation/306078/#.

224 *derived from the shout of "halloo":* Baron, *Always On*, 174.

225 *"Technology is neither good nor bad":* Melvin Kranzberg, "Technology and History:
'Kranzberg's Laws,'" *Technology and Culture* 27, no. 3 (July 1986): 544–60.

225 *"This binds together by a vital cord":* Briggs and Maverick, *The Story of the Telegraph*, 22.

225 *the "aerial" man . . . mobile phones:* Langdon Winner, "Sow's Ears from Silk Purses,"
in *Technological Visions: Hopes and Fears That Shape New Technologies*, eds.
Marita Sturken, Douglas Thomas, and Sandra Ball-Rokeach (Philadelphia: Temple
University Press, 2004), 34, 36.

227 *"The Strength of Weak Ties":* Mark S. Granovetter, "The Strength of Weak Ties,"
American Journal of Sociology 78, no. 6 (May 1973): 1360–80, accessed March 26,
2013, sociology.stanford.edu/people/mgranovetter/documents/granstrengthweak
ties.pdf; Mark Granovetter, *Getting a Job: A Study of Contacts and Careers* (Chicago: University of Chicago Press, 1995), 10–22, 51–53.

229 *what Malcolm Gladwell called connectors:* Malcolm Gladwell, *The Tipping Point: How Little Things Can Make a Big Difference* (New York: Little, Brown, 2000), 38–41.

230 *Peter Diamandis, the head of the X Prize Foundation:* Peter H. Diamandis, "Instant Gratification," in *Is the Internet Changing the Way You Think?: The Net's Impact on Our Minds and Future,* ed. John Brockman (New York: HarperCollins, 2011), 214.

231 *Facebook's news feed analyzes:* Eli Pariser, *The Filter Bubble: What the Internet Is Hiding from You* (New York, Penguin, 2011), 37–38, 217–43.

231 *people who are heavily socially active online:* Lee Rainie and Barry Wellman, *Networked: The New Social Operating System* (Cambridge, MA: MIT Press, 2012), Kindle edition.

232 *Consider the case of Maureen Evans:* Some of my writing here appeared in "Clive Thompson in Praise of Obscurity," *Wired,* February 2010, accessed March 26, 2013, www.wired.com/magazine/2010/01/st_thompson_obscurity/. In this book I use quotes from my interview with Evans that have slightly different phrasing than those in *Wired.* The story about Evans in *The New York Times* is Lawrence Downes, "Take 1 Recipe, Mince, Reduce, Serve," April 29, 2009, accessed March 26, 2013, www.nytimes.com/2009/04/22/dining/22twit.html.

235 *"they can't count as 'followers' in any meaningful sense":* Anil Dash, "Nobody Has a Million Twitter Followers," *Anil Dash: A Blog About Making Culture,* January 5, 2010, accessed March 26, 2013, dashes.com/anil/2010/01/nobody-has-a-million -twitter-followers.html.

236 *"If you give me six lines":* The *Oxford Dictionary of Quotations,* ed. Elizabeth M. Knowles (New York: Oxford University Press, 2001), 627.

237 *something shady about having different sides:* David Kirkpatrick, *The Facebook Effect: The Inside Story of the Company That Is Connecting the World* (New York: Simon and Schuster, 2010), 199.

237 *"We do not show ourselves":* James is quoted in Erving Goffman, *The Presentation of Self in Everyday Life* (New York: Anchor Books, 1990), 48–49.

237 *once famously said:* Richard Esguerra, "Google CEO Eric Schmidt Dismisses the Importance of Privacy," Electronic Frontier Foundation, December 10, 2009, accessed March 26, 2013, www.eff.org/deeplinks/2009/12/google-ceo-eric-schmidt -dismisses-privacy.

238 *an "omniopticon":* Nathan Jurgenson, "Film Review: Ondi Timoner (2009) *We Live in Public,*" *Surveillance & Society* 8, no. 3 (2010): 374–78.

238 *micro-celebrity:* Theresa Senft, *Camgirls: Celebrity & Community in the Age of Social Networks* (New York: Peter Lang, 2008), 25.

238 *People have been neurotically managing:* Goffman, *Presentation of Self,* 39, 47, 208, 236–37.

240 *cleverly found ways to wrest back some control:* danah boyd and Alice E. Marwick, "Social Privacy in Networked Publics: Teens' Attitudes, Practices, and Strategies," paper presented at A Decade in Internet Time: Symposium on the Dynamics of the Internet and Society, September 2011, accessed March 26, 2013, papers.ssrn.com/ sol3/papers.cfm?abstract_id=1925128; danah boyd, "Dear Voyeur, Meet Flâneur . . . Sincerely, Social Media," *Surveillance & Society* 8, no. 4 (2011): 505–07. Some of my writing here appeared in "Clive Thompson on Secret Messages in the Digital Age,"

Wired, February 2011, accessed March 26, 2013, www.wired.com/magazine/2011/01/st_thompson_secretmessages/.

241 *consider the example of Drop.io:* I described Drop.io's success with artificial forgetting in my column "Clive Thompson on Remembering Not to Remember in an Age of Unlimited Memory."

242 *"reputation bankruptcy":* Jonathan Zittrain, "Reputation Bankruptcy," *The Future of the Internet—and How to Stop It* (blog), September 7, 2010, accessed March 26, 2013, futureoftheinternet.org/reputation-bankruptcy.

242 *the European Commission has proposed:* Michael Venables, "The EU's 'Right To Be Forgotten': What Data Protections Are We Missing in the US?" *Forbes*, March 8, 2013, accessed March 26, 2013, www.forbes.com/sites/michaelvenables/2013/03/08/the-ecs-right-to-be-forgotten-proposal-in-the-u-s/.

243 *measure and broadcast crude measures of brain-wave activity:* Andrew T. Campbell et al., "NeuroPhone: Brain–Mobile Phone Interface Using a Wireless EEG Headset," *Proceedings of the Second ACM SIGCOMM Workshop on Networking, Systems, and Applications on Mobile Handhelds* (2010), 38, accessed March 26, 2013, sensorlab.cs.dartmouth.edu/pubs/neurophone.pdf.

Chapter 9: The Connected Society

245 *they don't have a very good reputation:* A concise survey of the libels against the post-'90s is Lara Farrar, "How Will China's Tech-Savvy, Post-90s Generation Shape the Nation?" CNN Tech, July 18, accessed March 26, 2013, articles.cnn.com/2010-07-18/tech/china.post90s.generation_1_internet-analyst-wild-card-generation-tencent?_s=PM:TECH. The description of "exploded head" hairstyles comes via the Chinese blogger Han Han: David Wertime, "Translation: Han Han Says Post-'90s Chinese the 'Masters of Tomorrow,' and Today," *Tea Leaf Nation*, July 6, 2012, accessed March 26, 2013, www.tealeafnation.com/2012/07/translation-han-han-says-post-90s-chinese-the-masters-of-tomorrow-and-today/.

245 *one of the most successful environmental protests ever:* Some stories detailing the Shifang protest include Tania Branigan, "Chinese City Cancels Copper Plant Project After Protests," *The Guardian*, July 3, 2012, accessed May 30, 2013, http:// www.guardian.co.uk/world/2012/jul/03/chinese-cancels-copper-plant-protests; Clifford Coonan, "The Day People Power Took On the Might Of China—and Won; Environmental Protest Forces U-turn on Plan to Build a Factory Despite Police Crackdown," *The Independent* (UK), July 4, 2012, accessed May 30, 2013, http://www.independent.co.uk/news/world/asia/the-day-people-power-took-on-the-might-of-china--and-won-7906813.html. Keith Bradsher, "Bolder Protests against Pollution Win Project's Defeat in China," *The New York Times*, July 4, 2012, accessed March 24, 2103, www.nytimes.com/2012/07/05/world/asia/chinese-officials-cancel-plant-project-amid-protests.html; Mark McDonald, "Taking It to the Street in China," *The New York Times*, July 29, 2012, accessed March 26, 2013, rendezvous.blogs.nytimes.com/2012/07/29/taking-it-to-the-street-in-china/; Ben Blanchard, "China Stops Copper Plant, Frees 21 after Protests," Reuters, July 4, 2012, accessed March 26, 2013, uk.reuters.com/article/2012/07/04/us-china-pollution-protest-idUKBRE86205C20120704; Choi Chi-yuk, "Post-90s Generation Speaks Up; Young People in Sichuan Led the Recent Opposition to a Heavy-Metal Plant, Causing Some to Hope They'll Fight for Other

Causes, Including Greater Democracy," *South China Morning Post*, July 14, 2012; Brian Spegele, "Planned China Metals Plant Scrapped," *The Wall Street Journal*, July 3, 2012, accessed March 24, 2013, online.wsj.com/article/SB10001424052702304211 804577504101311079594.html; Alia, "Traces on Weibo: How a NIMBY Protest Turned Violent in a Small Sichuan City?" *Offbeat China*, July 2, 2012, accessed March 26, 2013, offbeatchina.com/traces-on-weibo-how-a-nimby-protest-turned-violent-in -a-small-sichuan-city; Emily Calvert, "The Significance of Shifang: Environmental Activism in China," *China Elections & Governance* (blog), July 19, 2012, accessed March 26, 2014, chinaelectionsblog.net/?p=20482.

246 *rivers run in different colors:* Wendy Qian, "Han Han: The Liberation of Shifang," *China Digital Times*, July 3, 2012, accessed March 26, 2103, chinadigitaltimes .net/2012/07/han-han-the-release-shifang/.

247 *This action backfired badly:* Anne Henochowicz, "Shifang: A Study in Contrasts," *China Digital Times*, July 6, 2012, accessed March 24, 2013, chinadigitaltimes .net/2012/07/shifang-a-study-contrasts/; Alia, "Traces on Weibo"; Jing Gao, "Face off in Shifang—Photos of China's Largest and Bloodiest NIMBY Protest in Recent History," *Ministry of Tofu* (blog), accessed March 24, 2013, www.ministryoftofu .com/2012/07/faceoff-in-shifang-photos-of-chinas-largest-and-bloodiest-nimby -protest-in-recent-history/.

248 *a stunning 5.25 million postings:* Qian Gang, "China's Malformed Media Sphere," China Media Project, July 11, 2012, accessed March 26, 2013, cmp.hku.hk /2012/07/11/25293/.

248 *Shifang's police issued a warning:* Anne Henochowicz, "Shifang: A Study in Contrasts."

248 *"The information and pictures shared through Weibo":* Dexter Roberts, "On China's Web, Green Activists Grow Bolder," *Businessweek*, July 20, 2012, accessed March 26, 2013, www.businessweek.com/articles/2012-07-20/chinese-protesters -see-greater-online-freedom-over-environment.

248 *as one Weibo user noted:* Anne Henochowicz, "Netizen Voices: Hu Xijin on Shifang," *China Digital Times*, July 9, 2012, accessed March 26, 2013, chinadigitaltimes .net/2012/07/netizen-voices-hu-xijin-shifang/.

248 *The students were exultant:* Mark McDonald, "A Violent New Tremor in China's Heartland," *The New York Times*, July 4, 2012, accessed March 26, 2013, rendez vous.blogs.nytimes.com/2012/07/04/a-violent-new-tremor-in-chinas-heartland/; "Han Han Praises Young Protesters in Sichuan," *Want China Times*, July 6, 2012, accessed March 26, 2013, www.wantchinatimes.com/news-subclass-cnt.aspx?cid=1 103&MainCatID=&id=20120706000007.

249 *"It is a mystery to me":* Eisenstein, *The Printing Press as an Agent of Change*, 306.

249 *have long credited samizdat publishing:* Evgeny Morozov, *The Net Delusion: The Dark Side of Internet Freedom* (New York: PublicAffairs, 2012), xi–xii, 7, 34–36, 48–50, 64.

249 *"The Goliath of totalitarian control":* "Reagan: Take Risk on Arms Control," *Pittsburgh Post-Gazette*, June 14, 1989, accessed March 26, 2013, news.google.com/ne wspapers?nid=1129&dat=19890614&id=OelRAAAAIBAJ&sjid=zm4DAAAAIBAJ &pg=4965,3945114.

249 *as Hillary Clinton more recently proclaimed:* Hillary Rodham Clinton, "Remarks," speech delivered at the Conference on Internet Freedom, The Hague, December 8, 2011, accessed March 26, 2013, www.state.gov/secretary/rm/2011/12/178511.htm.

250 *8,709 per year:* Cary Huang, "Leaders Lost for Words to Describe and Address Cause of Social Strife," *South China Morning Post*, July 19, 2012, accessed March 26, 2013, www.scmp.com/article/727931/leaders-lost-words-describe-and-address-cause-so cial-strife.

251 *sociologist Hubert O'Gorman got interested in this puzzle:* Hubert J. O'Gorman, "Pluralistic Ignorance and White Estimates of White Support for Racial Segregation," *The Public Opinion Quarterly* 39, no. 3 (Autumn 1975): 313–30.

253 *a 1980 survey:* D. Garth Taylor, "Pluralistic Ignorance and the Spiral of Silence: A Formal Analysis," *The Public Opinion Quarterly* 46, no. 3 (Autumn 1982): 325.

253 *In corporations, board members might all privately notice:* James D. Westphal and Michael K. Bednar, "Pluralistic Ignorance in Corporate Boards and Firms' Strategic Persistence in Response to Low Firm Performance," *Administrative Science Quarterly* 50, no. 2 (June 2005): 262–98.

253 *university campuses are hotbeds of pluralistic ignorance:* Tracy A. Lambert, Arnold S. Kahn, and Kevin J. Apple, "Pluralistic Ignorance and Hooking Up," *The Journal of Sex Research* 40, no. 2 (May 2003): 129–33.

253 *Pluralistic ignorance is an information problem:* Andrew K. Woods, "These Revolutions Are Not All Twitter," *The New York Times*, February 1, 2011, accessed March 26, 2013, www.nytimes.com/2011/02/02/opinion/02iht-edwoods02.html.

253 *how U.S. campus officials started fighting binge drinking:* William DeJong and Jeff Linkenbach, "Telling It Like It Is: Using Social Norms Marketing Campaigns to Reduce Student Drinking," *American Association for Higher Education Bulletin* 52, no. 4 (1999): 13–16.

254 *why Egyptians wanted to revolt:* Some materials documenting the lead-up to the Egyptian uprising: David Wolman, *The Instigators* (New York: The Atavist, 2011); Anand Gopal, "Egypt's Cauldron of Revolt," *Foreign Policy*, February 16, 2011, accessed March 26, 2013, www.foreignpolicy.com/articles/2011/02/16/egypt_s_cauldron_of_ revolt; Michael Slackman, "Labor Protests Test Egypt's Government," *The New York Times*, April 28, 2010, accessed March 26, 2013, www.nytimes.com/2010/04/29/ world/middleeast/29egypt.html; Michael Slackman, "Egypt Concedes to Resistance on Privatization Push," *The New York Times*, June 27, 2010, accessed March 26, 2013, www.nytimes.com/2010/06/28/world/middleeast/28egypt.html; Will Kelleher, "Egypt: Food Security Is Still the Issue," *Think Africa Press*, February 9, 2012, accessed March 26, 2013, thinkafricapress.com/egypt/food-security-most-pressing -issue-new-government; "Long Awaited Egypt Minimum Wage Sparks Discontent," Agence France-Presse, November 18, 2010, http://maannews.net/eng/ViewDetails .aspx?ID=334131; Maryam Ishani, "The Hopeful Network," *Foreign Policy*, February 7, 2011, accessed March 26, 2013, www.foreignpolicy.com/articles/2011/02/07/ the_hopeful_network; David D. Kirkpatrick and David E. Sanger, "A Tunisian- Egyptian Link That Shook Arab History," *The New York Times*, February 13, 2011, accessed March 26, 2013, www.nytimes.com/2011/02/14/world/middleeast/14egypt -tunisia-protests.html; Jillian York, "The Arab Digital Vanguard: How a Decade of Blogging Contributed to a Year of Revolution," *Georgetown Journal of International Affairs* 13, no. 1 (Winter/Spring 2012): 33–42, accessed March 26, 2013, jilliancyork .com/wp-content/uploads/2012/02/33-42-FORUM-York.pdf.

255 *"Most of their marches and protests":* Sahar Khamis and Katherine Vaughn, "Cyberactivism in the Egyptian Revolution: How Civic Engagement and Citizen Journalism

Tilted the Balance," *Arab Media & Society* 14 (Summer 2011), accessed March 26, 2013, www.arabmediasociety.com/index.php?article=769.

255 *"Fear was embodied":* Wael Ghonim, *Revolution 2.0: The Power of the People Is Greater Than the People in Power: A Memoir* (New York: Houghton Mifflin Harcourt, 2012), Kindle edition.

256 *a collective action problem:* Zeynep Tufekci, "New Media and the People-Powered Uprisings," *MIT Technology Review*, August 30, 2011, accessed March 26, 2013, www.technologyreview.com/view/425280/new-media-and-the-people-powered -uprisings/.

256 *an event in June 2010 helped break that collective action problem:* The reporting on Ghonim's Facebook page derives primarily from *Revolution 2.0*, particularly the chapters "Kullena Khaled Said," "Online and in the Streets," "A Preannounced Revolution," and "January 25, 2011."

258 *about 850 civilians died:* Jess Smee, "The World from Berlin: 'The Egyptian Revolution Is at Risk'," *Spiegel Online*, June 04, 2012, accessed March 26, 2013, www .spiegel.de/international/german-media-questions-verdict-against-former-egyptian -president-mubarak-a-836837.html.

258 *Egypt's soccer fans, the Ultras:* This analysis of the Ultras comes from Lucie Ryzova, "The Battle of Muhammad Mahmud Street: Teargas, Hair Gel, and Tramadol," *Jadaliyya*, November 28, 2011, accessed March 26, 2013, www.jadaliyya.com/pages/ index/3312/the-battle-of-muhammad-mahmud-street_teargas-hair; "Meet Egypt's Ultras: Not Your Usual Soccer Fans," *National Post*, February 2, 2012, accessed March 26, 2013, sports.nationalpost.com/2012/02/02/meet-egypts -ultras-not-your-usual-soccer-fans/; James Montague, "Egypt's Revolutionary Soccer Ultras: How Football Fans Toppled Mubarak," CNN, June 29, 2011, accessed March 26, 2013, edition.cnn.com/2011/SPORT/football/06/29/football.ultras.zamalek.ahly/index. html; and a presentation I attended by the Egyptian activist Alaa Abd El-Fattah at the Personal Democracy Forum in New York, June 6, 2011.

258 *Tufekci and her research team:* Zeynep Tufekci and Christopher Wilson, "Social Media and the Decision to Participate in Political Protest: Observations from Tahrir Square," *Journal of Communication* 62, no. 2 (April 2012): 363–79.

258 *"Speaking, writing, and thinking":* Michael Warner, *Publics and Counterpublics* (Brooklyn, NY: Zone Books, 2005), 69.

259 *"The telegraph enabled people":* James Gleick, *The Information: A History, A Theory, A Flood* (New York: Pantheon, 2011), 147.

259 *a stern YouTube warning:* Courtney C. Radsch, *Unveiling the Revolutionaries: Cyberactivism and the Role of Women in the Arab Uprisings* (James A. Baker III Institute for Public Policy, May 18, 2012), accessed March 26, 2013, bakerinstitute.org/ publications/ITP-pub-CyberactivismAndWomen-051712.pdf.

259 *didn't think anyone would step away from the screen:* Ghonim, *Revolution 2.0*, Kindle edition.

260 *As Clay Shirky documents in* Cognitive Surplus*:* Shirky, *Cognitive Surplus*, 31–37.

260 *As the British media critic Charlie Beckett writes:* Charlie Beckett, "After Tunisia and Egypt: Towards a New Typology of Media and Networked Political Change," *Polis* (blog), February 11, 2011, accessed March 26, 2013, blogs.lse.ac.uk/polis/2011/ 02/11/after-tunisia-and-egypt-towards-a-new-typology-of-media-and-networked -political-change/.

261 *the more civically active they are offline:* Lee Rainie, Kristen Purcell, and Aaron Smith, *The Social Side of the Internet* (Pew Internet & American Life Project, January 18, 2011), 2, accessed March 26, 2013, www.pewinternet.org/~/media//Files/Reports/2011/PIP_Social_Side_of_the_Internet.pdf.

261 *selective exposure:* Farhad Manjoo, *True Enough: Learning to Live in a Post-Fact Society* (New York: John Wiley & Sons, 2011), 30.

262 *very little political agreement even among friends:* Lee Rainie and Aaron Smith, *Social Networking Sites and Politics* (Pew Internet & American Life Project, March 12, 2012), 2, accessed March 26, 2013, pewinternet.org/~/media//Files/Reports/2012/PIP_SNS_and_politics.pdf.

262 *on Twitter users who displayed "clear political preference":* Jisun An, Meeyoung Cha, Krishna Gummadi, and Jon Crowcroft, "Media Landscape in Twitter: A World of New Conventions and Political Diversity," presented at the Fifth International Association for the Advancement of Artificial Intelligence Conference on Weblogs and Social Media (2011), accessed March 24, 2013, http://www.cl.cam.ac.uk/~jac22/out/twitter-diverse.pdf.

263 *several studies on young people's online discussions:* Joseph Kahne, Ellen Middaugh, Nam-Jin Lee, and Jessica T. Feezell, "Youth Online Activity and Exposure to Diverse Perspectives," *New Media Society* 14, no. 3 (2011): 492–512; Joseph Kahne, Nam-Jin Lee, and Jessica T. Feezell, "Digital Media Literacy Education and Online Civic and Political Participation," *International Journal of Communication* 6 (2012): 1–24; Joseph Kahne, Nam-Jin Lee, and Jessica T. Feezell, "The Civic and Political Significance of Online Participatory Cultures among Youth Transitioning to Adulthood," *Journal of Information Technology & Politics* 10, no. 1 (2013): 1–20; and a personal interview with Kahne.

264 *a mere 2.9 percent of the time:* R. I. M. Dunbar, Anna Marriott, and N. D. C. Duncan, "Human Conversational Behavior," *Human Nature* 8, no. 3 (1997): 236. Actually, the percentage of time spent talking about politics is even smaller, because the 2.9 percent figure includes both politics and religion.

264 *seeking justice for a black teenager:* Kelly McBride, "Trayvon Martin Story Reveals New Tools of Media Power, Justice," Poynter.org, March 23, 2012, accessed March 26, 2013, www.poynter.org/latest-news/making-sense-of-news/167660/trayvon-martin-story-a-study-in-the-new-tools-of-media-power-justice/; Miranda Leitsinger, "How One Man Helped Spark Online Protest in Trayvon Martin Case," NBC News, March 29, 2012, accessed March 26, 2013, usnews.nbcnews.com/_news/2012/03/29/10907662-how-one-man-helped-spark-online-protest-in-trayvon-martin-case; An Xiao Mina, "A Tale of Two Memes: the Powerful Connection Between Trayvon Martin and Chen Guangcheng," *The Atlantic*, July 12, 2012, accessed March 26, 2013, www.theatlantic.com/technology/archive/12/07/a-tale-of-two-memes-the-powerful-connection-between-trayvon-martin-and-chen-guangcheng/259604/.

265 *The Haitian earthquake of 2010:* Jessica Heinzelman and Carol Waters, *Crowdsourcing Crisis Information in Disaster-Affected Haiti* (United States Institute of Peace, October 2010); Francesca Garrett, "We Are the Volunteers of Mission 4636," *Ushahidi* blog, January 27, 2010, accessed March 26, 2013, blog.ushahidi.com/2010/01/27/mission-4636/.

266 *"When compared side by side"*: Heinzelman and Waters, *Crowdsourcing*, 2.

266 *slower-moving civic issues:* Silvia Vinas, "Colombia: #Yodigoaquiestoy, a Tool for De-nouncing Child Labor," trans. Thalia Rahme, *Global Voices*, July 22, 2013, accessed March 26, 2013, globalvoicesonline.org/2012/07/22/colombia-yodigoaquiestoy-a-tool-for-denouncing-child-labor/; Geoavalanche, accessed March 26, 2014, geo avalanche.org/incident/.

266 *"Having a real-time map"*: Patrick Philippe Meier, "Do 'Liberation Technologies' Change the Balance of Power Between Repressive States and Civil Society?" (PhD dissertation, Tufts University, 2011), 12.

266 *Steve Mann calls it "sousveillance"*: Some of the writing here previously appeared in my column "Clive Thompson on Establishing Rules in the Videocam Age," *Wired*, July 2011, accessed March 26, 2013, www.wired.com/magazine/2011/06/st_thomp son_videomonitoring/. Mann himself writes about the concept here: Steve Mann, Jason Nolan, and Barry Wellman, "Sousveillance: Inventing and Using Wearable Computing Devices for Data Collection in Surveillance Environments," *Surveillance & Society* 1, no. 3 (2003), 331–55, accessed March 26, 2013, library.queensu.ca/ojs/index.php/surveillance-and-society/article/view/3344/3306.

267 *found that about 50 percent documented the protests:* Tufekci and Wilson, "Social Media and the Decision to Participate," 14.

267 *"There is now a constant 'price tag'"*: "A New Chapter of People Power," *The Euro-pean*, May 3, 2012, accessed March 26, 2013, www.theeuropean-magazine .com/571-popovic-srdja/570-global-non-violent-activism.

268 *the "donkey blogger" protest:* My description of the protest and Pearce's findings de-rives from personal interviews with Pearce and Hajizada, as well as the following documents: Katy E. Pearce and Sarah Kendzior, "Networked Authoritarianism and Social Media in Azerbaijan," *Journal of Communication* 62, no. 2 (April 2012): 283–98; Brian Whitmore and Anna Zamejc, "Azeri Bloggers Receive Prison Sen-tences for 'Hooliganism,'" *Radio Free Europe Radio Liberty*, last updated November 11, 2009, accessed March 26, 2013, www.rferl.org/content/Azerbaijan_Bloggers_Get_TwoYear_Jail_Sentences/1874853.html; Lisa Margonelli, "Update: Azerbaijan's 'Donkey Bloggers' Get 2 Years in Prison," *The Atlantic*, November 11, 2009, ac-cessed March 26, 2013, www.theatlantic.com/technology/archive/2009/11/update -azerbaijans-donkey-bloggers-get-2-years-in-prison/30000/.

270 *"networked authoritarianism"*: Rebecca MacKinnon, *Consent of the Networked: The Worldwide Struggle for Internet Freedom* (New York: Basic Books, 2012), Kin-dle edition.

270 *"Before the advent of social media"*: Morozov, *Net Delusion*, 156.

270 *"to be preserved forever"*: Mayer-Schönberger, *Delete*, 103.

270 *the proregime* Raja News *printed thirty-eight photos:* Morozov, *Net Delusion*, 10.

271 *a troubling side effect to sousveillance:* John Villasenor, *Recording Everything: Digi-tal Storage as an Enabler of Authoritarian Governments* (Brookings Institution, December 14, 2011), 8.

271 *a documentary crew in 2005 showed Beijing university students:* MacKinnon, *Consent of the Networked*, Kindle edition.

272 *they delisted the page:* Ghonim, *Revolution 2.0*, Kindle edition.

273 *In 2004, Yahoo! handed over to the Communist Party:* I reported on the Shi Tao case

in "Google's China Problem (and China's Google Problem)," *The New York Times Magazine*, April 23, 2006, accessed March 26, 2013, www.nytimes.com/2006/04/23/magazine/23google.html.

273 *"On Apple's special store for the Chinese market":* MacKinnon, *Consent of the Networked*, Kindle edition.

273 *Western firms that sell to despotic regimes:* Thompson, "Google's China Problem"; Morozov, *Net Delusion*, 221; Paul Sonne and Steve Stecklow, "U.S. Products Help Block Mideast Web," *The Wall Street Journal*, March 27, 2011, accessed March 26, 2013, online.wsj.com/article/SB10001424052748704438104576219190417124226 .html; Jennifer Valentino-DeVries, Paul Sonne, and Nour Malas, "U.S. Firm Acknowledges Syria Uses Its Gear to Block Web," *The Wall Street Journal*, October 29, 2011, accessed March 26, 2013, online.wsj.com/article/SB10001424052970203687 504577001911398596328.html.

273 *the FBI violated the law thousands of times:* "Patterns of Misconduct: FBI Intelligence Violations from 2001–2008," Electronic Frontier Foundation, February 23, 2011, accessed March 26, 2013, www.eff.org/wp/patterns-misconduct-fbi-intelligence -violations.

273 *Amazon and Paypal cut off Wikileaks:* Rebecca MacKinnon, "WikiLeaks, Amazon and the New Threat to Internet Speech," CNN, December 3, 2010, accessed March 26, 2013, www.cnn.com/2010/OPINION/12/02/mackinnon.wikileaks .amazon/.

273 *"A Declaration of the Independence of Cyberspace":* John Perry Barlow, "A Declaration of the Independence of Cyberspace," February 8, 1996, accessed March 26, 2013, projects.eff.org/~barlow/Declaration-Final.html.

274 *The free, open-source Tor:* The Tor Project is online here: www.torproject.org/ (accessed March 26, 2013); other modes of encrypted communications are detailed in "Learn to Encrypt Your Internet Communications," Electronic Frontier Foundation, accessed March 26, 2013, ssd.eff.org/wire/protect/encrypt.

274 *Even sousveillance can be made safer:* Bryan Nunez, "ObscuraCam v1: A Mobile App for Visual Privacy," *WITNESS Blog*, June 29, 2011, accessed March 26, 2013, blog .witness.org/2011/06/obscuracam/; Deborah Netburn, "YouTube's New Face-Blurring Tool Designed to Protect Activists," *Los Angeles Times*, July 18, 2012, accessed March 26, 2013, articles.latimes.com/2012/jul/18/business/la-fi-tn-you tube-face-blurring-20120718.

274 *DIY social networks:* "Welcome to Crabgrass," Riseuplabs, accessed March 26, 2013, crabgrass.riseuplabs.org/; The Diaspora Project, accessed March 26, 2013, diasporaproject.org/.

274 *what the donkey blogger Hajizada wrote:* "I Spent 17 Months in a 3rd World Country (Horrible Conditions) Jail Because of a YouTube Video I Made. AMA," Reddit, July 30, 2012, accessed March 26, 2013, www.reddit.com/r/IAmA/comments/xdxco/i_ spent_17_months_in_a_3rd_world_country_horrible/.

275 *the Cute Cat Theory:* Ethan Zuckerman, "The Connection between Cute Cats and Web Censorship," *My Heart's in Accra* (blog), July 16, 2007, accessed March 26, 2013, www.ethanzuckerman.com/blog/2007/07/16/the-connection-between-cute -cats-and-web-censorship/; Ethan Zuckerman, "Cute Cat Theory: The China Corollary," *My Heart's in Accra*, December 3, 2007, accessed March 26, 2013, www .ethanzuckerman.com/blog/2007/12/03/cute-cat-theory-the-china-corollary/.

275 *"Cartoons Against Corruption":* "Why Is This Cartoonist Caged?" *The Hoot*, April 23, 2012, accessed March 26, 2013, www.thehoot.org/web/home/story.php?storyid =5881&mod=1&pg=1&sectionId=21; "In India the Enemies of Free Speech Find a 'Symbolic' Means to Attack Cartoonist Aseem Trivedi," Cartoonists Rights Network International, February 9, 2013, accessed March 26, 2013, www.cartoonistsrights .org/recent_developments_article.php?id=28; Cartoons Against Corruption, accessed March 26, 2013, cartoonsagainstcorruption.blogspot.com/.

276 *a new Magna Carta for the digital age:* Tim Berners-Lee, "Long Live the Web: A Call for Continued Open Standards and Neutrality," *Scientific American*, November 22, 2010, accessed March 26, 2013, www.scientificamerican.com/article.cfm?id=long -live-the-web.

276 *push for regulations:* Venables, "The EU's 'Right to Be Forgotten.'"

276 *When student activists pressured apparel companies . . . Global Network Initiative:* MacKinnon, *Consent of the Networked*, Kindle edition.

277 *As Cory Doctorow points out:* Cory Doctorow, "We Need a Serious Critique of Net Activism," *The Guardian*, January 25, 2011, accessed March 26, 2013, www .guardian.co.uk/technology/2011/jan/25/net-activism-delusion.

Chapter 10: Epilogue

279 *The* Jeopardy! *clue popped up on the screen:* Parts of my writing here on Watson appeared previously in my story "What Is I.B.M.'s Watson?" *The New York Times Magazine*, June 16, 2010, accessed March 26, 2013, www.nytimes.com/2010/06/20/ magazine/20Computer-t.html.

283 *Indeed, some clever experiments by Harvard's Teresa Amabile and others:* Teresa M. Amabile, "Brilliant but Cruel: Perceptions of Negative Evaluators," *Journal of Experimental Social Psychology* 19, no. 2 (March 1983): 146–56; Bryan Gibson and Elizabeth Oberlander, "Wanting to Appear Smart: Hypercriticism as an Indirect Impression Management Strategy," *Self and Identity* 7, no. 4 (2008): 380–92.

284 *"the smartest medical student we have ever had":* Joanna Stern, "IBM's Watson Supercomputer Gets Job as Oncologist at Memorial Sloan-Kettering Cancer Center," ABC News, March 22, 2012, accessed March 26, 2013, abcnews.go.com/ Technology/ibms-watson-supercomputer-job-memorial-sloan-kettering-cancer/story ?id=15979580#.UVQxTKt5MhM.

285 *A Pew study found that 22 percent of all TV watchers:* Aaron Smith and Jan Lauren Boyles, "The Rise of the 'Connected Viewer'" (Pew Internet & American Life Project, July 17, 2012), 2, accessed March 26, 2013, pewinternet.org/Reports/2012/ Connected-viewers.aspx.

285 *rebuslike short forms of expression:* I owe this idea to Crystal's *Txtng*, 39.

285 *Andy Hickl, an AI inventor:* Thompson, "What Is I.B.M.'s Watson?"

287 *As one of the commenters at the site explained:* "Reader's Comments: What Is IBM's Watson?" accessed March 26, 2013, community.nytimes.com/comments/www.ny times.com/2010/06/20/magazine/20Computer-t.html?sort=recommended.

288 *to a 2004 discussion-board thread:* "Favorite Movie Quotes," accessed March 26, 2013, www.clemsontalk.com/vb/archive/index.php/t-2926.html.

Index